*In Praise of
Practical Fertilizer*

In Praise of
Practical Fertilizer

Scenes from Chester
Township

JOHN BASKIN

W · W · NORTON & COMPANY

NEW YORK · LONDON

Most grateful acknowledgment to William C. Pendleton and the Ford Foundation, for interest and support.

Copyright © 1982 by John Baskin. *All rights reserved.* These essays have appeared in *The New York Times, The Washington Post, The Boston Globe, The Cincinnati Enquirer, Country Journal, Harper's Weekly, The Yale Review, American Preservation,* and *Ohio Magazine.* Published simultaneously in Canada by George J. McLeod Limited, Toronto. Printed in the United States of America.

Library of Congress Cataloging in Publication Data
Baskin, John.
    In praise of practical fertilizer.
        1. Chester region (Clinton County, Ohio)—
Social life and customs.   2. Chester region
(Clinton County, Ohio)—Description and travel.
I. Title.
F499.C39B28   977.1'765   82–2112
                                AACR2

ISBN 0-393-01563-7

W. W. Norton & Company, Inc. 500 Fifth Avenue, New York, N. Y. 10110
W. W. Norton & Company Ltd. 37 Great Russell Street, London WC1B 3NU

1 2 3 4 5 6 7 8 9 0

For Dan and Mary, Miss Button and Mr. Bobo
and, of course,
The Squire

I dream upon the opposing lights of the hour

—*Waiting*, Robert Frost

"Homer," he said, "we have seen many things in our time."

— Gurneyville milkman

Contents

Vantage Point   19

In a Poor Season   21

January   26

Sarah and Dora   40

Mr. Baker and the Weatherman   42

Provocative Vegetables   44

February   46

Groundhogs, in Defense of   48

Crime and Punishment, I   53

Mr. Baker Fusses into the Year   60

March   62

The News in the Crevices   67

The Late, Great New Burlington   70

Crime and Punishment, II   76

Mr. Robinson's Sober Music   80

April   82

An Essay on Taxes, Woodstoves, and the Underlying Frivolity of Systems (Both Human and Mechanical) and Some Thoughts Thereof: Or How Mr. Berger, Lawyer and Woodcutter, Takes on the Government in a Minor but Principled Skirmish   84

Syruping, to No One's Liking   90
A Dearth of Porches   93
On Hogs   95
Village Life, Hold the Consecration   97
Desire amidst the Clover   101
The Confounding Manifestations of Camping   103
Small Talk   105
Mortal Occasions   108
A Quiet Day with Geese-Watching   110
Gardening, I   112
Gardening, II   115
A Slight Sound at Evening (and Again at Morning)   117
The Weather in Gurneyville   119
Let Us Now Praise Famous Sharecroppers   126
Auctioneer's Song   129
The Sound of Ingenuity   133
Of Crowbar and Profanity   137
Chester Township Boys Go to War   141
In Praise of Practical Fertilizer   146
The Cold Boards of Morning   149
Mr. Baker Fusses into Summer   154
Passion among the Limas   156
Heat   158
Grazing Rites   162
Mildew   164
Requiem for a Hotel   165
Baseball on the Backlots   168
Ruminations on Cow Manure   172
Country Correspondence   175
Last Respects for Mr. Humphrey   178

Contents

On Storytelling   180
A Cornerstone of Pie   182
An Old Picture, in the Mind   185
Backfield Dreams   187
Conversation in a Graveyard   189
On History   194
First Frost   196
Home Is Where the Hearth Is   198
A Chance of Widely Scattered Contradictions   203
God's Winter, Man's Lot   206
Inching Up on Winter   211
First Snow, Baleful Mercury   213
A Christmas Memory   214

Preface

IN *Moby-Dick*, when the narrator tells the reader that Queequeq is a native of the island of Kokovoko, there is a splendid sentence that I've always remembered: *It is not down on any map; true places never are.* I was sorely tempted by that sentence when I came down to selecting the properly erudite epigram for these essays, the lovely, final, often presumptuous act in writing a book.

I have been arrogantly off again, building little universes. A few years ago, I built a place called New Burlington. It, like Chester Township, is not on many maps. But Melville had large lusts; mine are small. He was after the whale; this is a book of minnows.

Chester Township does exist, of course. A reader might ring up The Squire, the beekeeper, or the syrup-maker (although they are not listed so in the telephone book), and ask if they have been attended to properly. That fidelity exists, although I'm not sure it matters much. For me, the township exists more in the mind's eye, that peculiar and private geography by which I usurp territory for myself.

I cannot even tell anyone exactly where the actual township lies. It is somewhat south of Xenia, a few miles northwest of Wilmington, slightly to the left of the Civil War, and a bit right of the moonwalk. The Squire, a more precise and historically oriented man, tells me boundaries in the names of counties, other townships. When I press him for translation, he impatiently tells me that Chester Township is roughly bounded by New Burlington to the north (as though it still existed), the new reservoir to the west, Gurneyville to the east, and Weldon Heller's old house on the south, although the township line went somewhere through his yard and no one was ever quite sure where the Heller kids were supposed to go to school.

Our township consists of 20,123 acres, upon which may be found 231 farms, fifty residences, one village, one school, three churches, and four cemeteries. It is an old township, one of three which made up Clinton County in 1810. It has suffered the infractions of the modern world as much as most, survived more intact than many. The lake took New Burlington, leaving the township with one village, and the interstate divided it, cutting a swath through its best farmland, including sixteen acres of The Squire's place.

He accepted that with the equanimity he reserved for hail on the new corn, politics, and most of the other improprieties of man and nature. The Squire possessed a long view of history; the earth had its sacredness but highway engineers had eminent domain. On a bleak winter day he sat in his kitchen and watched a bulldozer unloaded on ground his great-grandfather had begun to farm in 1853 after leaving Wales. "Well," he said to Vivian, "there they go. . . ."

The Squire knew what was occurring. The township was going, too. Not all at once but like most things, one fragment at a time. *Change* was occurring and The Squire had himself prepared for it, at least as well as he knew how. If he lost sixteen acres of old William Williams's topsoil, then was there something he might gain from the interstate? In time, he learned that he could plant his corn by the first load of water-melons headed north on the interstate in the spring. He learned that when traffic noise was loudest, the weather was about to change. And he learned that he could, with half the former difficulty, drive over to see his children who were attending the state university.

The Squire's neighbor out on Route 73, Eddie Shidaker, watched the engineers drive stakes into his yard and decided it was time to die. And he did. The Squire watched other farm-ers do the same thing. That was not *his* way. He wanted to adapt. He didn't feel particularly forced to. It was just the way he was. Change came, he said, with hosannas and hoodoos, and you better be prepared for both. Be content with small gains, cut your losses. Farming, The Squire thought, should have prepared us all better.

In his time, which was not so far distant, Chester Town-ship had good trustees, a good school system, good roads, good farms. When asked where they were from, local residents said, "Chester Township." It was a place in the world. But gradu-ally the righteous definition *township* was lost. The schools consolidated, the interstate came, families dispersed, the old village disappeared. The township became a forgotten bound-ary, a "past figure of government speech," The Squire put it.

He still told people he was from Chester Township (not Clinton County, nor even Wilmington) and so did his neigh-bor across the road, the beekeeper, who was a recent arrival. (He and his wife, Mary, had come up from Cincinnati to pick berries one summer afternoon and just stayed.) The bee-keeper, although he was thirty years The Squire's junior, had some old-fashioned ways about him. One of the things he liked to do occasionally was have The Squire bring over his old slides of how the neighborhood once looked, and how The

Squire and his father had started farming together, and even the slides of the engineers building the interstate through his cornfields. He listened to The Squire's running commentary as though one day he might take over the slides and give the township lecture himself.

The Squire said my interest in Chester Township was likely because I lived in the old Lane house, and it was in the center of the township, as though geography made its own demands, which, of course, it does. He showed up there at odd times, always welcome, and charmed his way past the old labrador who was always a sucker for his overalls, wealthy in mysterious scents from the east, to bring me an old book, newspaper clipping, or some story he thought I might like.

The township continued to exist among us, although it was as much a mythological landscape as a place of real boundaries and we were the mythfits who inhabited both. It was a country of no large events, some amount of decorum, the sharing of work and experience and, let us say, *manners.*

Nonfiction is a craft, not an art, or perhaps it is one of the natural sciences, like geology, and Chester Township one of the compressed layers of time. We are no longer inhabitants of something so paleotechnic as the township, and we know it. We are one-worlders now, and we have box numbers and zip codes. The beekeeper drives fifty miles to work in a distant city and The Squire has a son who is an opera singer in New York. And although I have lived in Chester Township for most of a decade, I am as homeless and transient as ever. We all still fear eminent domain, wishing instead such a principle for ourselves. These essays are merely one more place, or a glimpse of a place in time. Chester Township may not be my place but like New Burlington before it, it has been my place for a time, and for a time that will suffice.

In Praise of
Practical Fertilizer

Vantage Point

THIS HOUSE is my vantage point. It is a two-story farmhouse in the middle of two hundred acres of cornfields. Lester and Vera Lane built it in 1924, and the neighbors say it was quite a place. There were hardwood floors, a full basement, and on either side of the big living-room fireplace, glass-doored bookcases containing Shakespeare and Macauley's essays. Mrs. Bailey down the road said she'd never seen a closet just for brooms before.

The children were not as impressed as the neighbors. "We had that big house," said one of them, "and nothing in it." Finally, in the generosity of good markets just before that hair shirt of a decade to come, Vera bought a good carpet for the living room. In Xenia, there was a man also named Lester Lane, and the bill went to his house where it was opened by his wife, Gertie. "Lordy," she said delivering the bill to Vera, "that must be *some* carpet."

Vera was from a well-off family. Her father had one of the first two automobiles in Clinton County. On a Sunday, he took his family driving and ran into the second car. Lester's father was a night watchman at the casting factory. He came down the stairs at four o'clock every afternoon with his shoes in his hands and sleep on his face. Lester considered his family poverty-stricken. This image informed him. He was determined to "get-on." Early photographs of him show a handsome but grave young man. His credo in the school yearbook was "serious, solemn, and studious."

When he built this house, it cost only slightly less than the surrounding cornfields. In the severe weather of the thirties, he almost lost both. To save them, he went to work in the federal land bank program, although he never had any use for Democrats. But neither did anyone else in the neighborhood. Vergo Mitchner wouldn't carry a Roosevelt dime in his pockets.

Once, on a trip to Washington with the farm bureau, the group drove down Pennsylvania Avenue, past the White House, and Lester got out and asked one of the guards if he might see Mr. Roosevelt for a moment. Presumably, he and the president would have soon put things to rights.

Lester beat the depression, though. It is possible he even enjoyed it a bit, like an affliction one whips, then feels stronger for it. The maples his sons planted around the house matured, and so did the sons, although one fell under suspicion because he became a Unitarian minister.

"A Unitarian," sniffed one old Methodist, "is someone who believes Jesus Christ was the illegitimate son of Joseph. Now what would *your* mother think of that?"

When Ed graduated from Wilmington College and went off to the seminary, the big farmhouse was nearly empty. "I feel a bit too old, and finished," Vera wrote in her diary. She liked the minister who told her heaven was a place where "parents are young and children small."

She was not "too old" until well into her eighties, however. Then she became sick and needed the strict and professional observances of a nursing home. Lester went with her. They left the house as though they were going for a Sunday drive.

By the time I met Lester, he, too, was ill. He sat in a chair in the winter sunlight and talked about his trip back from Nebraska in a covered wagon, when he was ten. He was nearly deaf, and it was difficult for us.

At the farmhouse, I packed things away, moved some of the furniture upstairs. There were notes and records everywhere, in Lester's gradually failing handwriting. There were even notes scribbled on the holding tank for the cistern. This was partly his penchant for order but it seemed, too, a way for him to fight with his faculties. He fought the Depression, the land, the image of his family, and now he was fighting his memory. It was just one more fight, and as fierce as all the rest.

He saved everything, too. The house was filled with the accumulated equipage of over fifty years in one place. It was like moving into a museum. He'd even saved the gallstones from the operation on old Major, his best plowhorse.

The last time I saw him he was past ninety and his son Howard drove him by one Sunday afternoon. I was in the yard, pruning the shrubbery. He seemed pleased, and I certainly was. I felt a bit like one of the troops being reviewed by the old general. I was pleased I'd been working. That has been several years and even now, at certain odd times, I expect him to walk in and ask me why I'm not getting more done.

In a Poor Season

To the untrained eye in a poor season, the country here is all isolation. The touring eye has its troubles finding the local points of reference. The roads wander off on errands purely of their own, as any good country road should do, and it is this fact that always keeps me awake and interested on the few miles into town, although I regard the interstate to Columbus as the road God has picked out for irretrievable sinners to travel endlessly back and forth when Judgment gets here (as it seems to threaten these days, with ever more certainty).

The landmarks and the country are subtler here and it is still possible to drive at least half a mile between houses, and sometimes more, if you pick one of the little unpaved lanes still around. The first time I saw any of it, leaving the ornate mountains of western North Carolina behind, I looked it over and only the fact of having worn out my welcome south kept me north. It was the old, not entirely unsordid story of north toward welfare.

Once or twice, friends from New York have come out to visit and they like to sit in the yard, look philosophically into the distance, and muse about tranquility and clean air. I privately suspect, however, that we get in time a good bit of Cincinnati and Dayton in our air, and wait for the wind to change from Mr. Hewitt's hogbarn, which it does infrequently, prompting one visitor, stepping out on the porch for a morning shot of country air, to step back inside and quote a line from an old Victor Mature movie: "I think the wind is coming from the elephant yards."

I don't mind the windborne smell of Mr. Hewitt's commerce very much, and suspect things even out when the winds blow back his way, carrying with it the high odor of an overripe manuscript. As for tranquility, visitors usually manage to miss haying season, or events such as old Clyde, the bucksheep, chasing off a whole carload of Seventh-Day Adventists (the side pasture was thereby sowed in a rain of religious pamphlets, which proved to be nongerminating as Clyde later ate them all).

The sounds across the way are pleasant, too: a disgruntled

cow sending out a muted bellow of barnyard complaint; a rooster in the early morning hours, presiding over daylight. Every time I hear him I think of a cartoon I once saw in which, as the rooster strides purposefully past, one hen turns admiringly to the others and says: "There he goes again, girls. Off to make the sun rise."

My father is one of the few visitors who pronounced, at first glance, this country beautiful and, admittedly, his sense of aesthetics came from seeing, finally, land good enough to farm instead of those undernourished Carolina fields which, in a neighbor's description, were "so poor a killdee had to pack a lunch to get from one side to the other."

My father likes this country best in late summer, when the cornfields are approaching their zenith, and to be set down in the middle of one would require bearers with machetes and a fix on the polestar to get back to civilization. I like it, too, at these times but partly because I can sit on the backsteps in my underwear. The house cat seems as partial to tall corn as any of us, for she is fond of wandering off in the fields, then sitting down and yowling plaintively for awhile. The reasons for this odd display are known only to her, as are many things about a cat, although perhaps she is merely determining her direction, by echo.

I'm sure that when my father looks at these fields he imagines his own cattle growing fat. I can see that, too, although I see Angus while he sees Holstein. I see Angus because they are not milked at 4:30 each morning. To this day, my predilection for rising and retiring date to our dairy herd, one which demanded the flinty routines of boot camp, and so I keep big-city hours here in the country, rising late and retiring late. Early to bed, early to rise, in my experience, makes a man mostly unquotable.

Today, the only way I could milk a cow at her appointed time would be to just stay up another hour or two, then go to bed afterward. I do get up early several times a year, provoked by the ecstacies of one or another nature writer, or perhaps by guilt, but I've yet to be much impressed, so I've cut out nature writers.

In late summer, the corn covers up the country somewhat in the manner snow does in January, growing over old machinery and a multitude of sins, to be revealed at harvest time as still with us, as though by thaw. At that time, the fields themselves are landmarks. A man a few roads over, a big corn farmer, took a vacation last year by renting a small plane and flying over Indiana and Illinois, getting a good look at *those* cornfields.

There're no landmarks, in the tourist sense, around this country. It isn't very historical, although one of my neighbors has suggested the state put a historical marker beside a rather singular pothole in front of his place. Nobody too famous has come out of it, either. A writer I know of was once told by a writer he knew, just after the first fellow was preparing to launch himself into publicity for a book he'd written, "Going out to be famous, huh? Well, you go on out there. I been famous once or twice and it don't hurt much, and it don't last long." Most of the neighborhood fame is like that. The Colletts, who raise corn on this place, get famous along about August. I've got another neighbor who gets famous in sugar-making time, and one next place over, an electrician, gets famous every time there's a power failure.

No one famous even slept here that I know of, although a New York publisher spent the night upstairs once. He seemed a bit haggard the next morning, even though I told him the mice were harmless and if he felt the bed moving in the night to just throw his shoe against the wall. I also told him the stairs creaked of their own accord like stairs do in all old hardwood houses, and there was no truth to the rumor that Lester Lane's spirit roamed about at night looking for a piece of baling wire to repair the manure spreader with. The publisher left just after breakfast, and I subsequently named the front bedroom after him.

It is not a country of monuments, and when I think of it, the pictures in my mind have to do with light and dark and the gradations between, in all the odd and quiet moments of a year's passing. I see myself living in a curiously shaped little section of country, bordered in each direction by one neighbor or another, landmarks themselves.

I think of them as something like posts in an imaginary fence around the vague ground of my neighborhood. Some of them are good, solid end and corner assemblies set in cement, some are pine dipped in creosote, and some are warped old locust. Which is how it is. I once worked in a place where everyone was chosen for his or her intelligence, kindness, courage, and so forth, and it was one of the most miserable jobs I've ever had.

About the nearest thing to a curiosity in the neighborhood is Tom McMillan's round barn, or Sharon Church where old Sam Ellis, who chopped a tree down on himself, is buried in two boxes. Or what is more likely a monument to curiosity, one or the other of us.

For my city acquaintances, I sometimes do a fair, if exaggerated, imitation of one or more of my neighbors, but the truth is, the traits I admire I have a hard time imitating. If eccentric qualities are what strike the eye of a satirist, then it's likely I'm a prime candidate. For how strange is it when a fellow thinks he needs several hundred acres of cornfields as writing space?

January

1.

January's a fishhook of a month—steel in a tender place. How do I put myself in such a place? Where's heat enough when every flame seems banked? It's a month whose major idea is insolvency. One must be in shape for January, have a little something put away. It's no beginning, that's for sure. It's the dead, unearthly-quiet heart of the year. Who'd pick January for a beginning? It's a month of endings, a graveyard of a month. I've a neighbor who calls it "the Great Earnestness." We cover up in January yet nothing's quite enough; we're *still* exposed. The cold sinks down, heavy as any failure. Where's the heart for it when the instinct is an animal instinct, to burrow, to lie underneath?

Literature uses heat as an image of purification but I'd pick cold. Even petty crime stops when there's cold enough, but heat's an inspiration. January's a month for the intellect; there's no passion left. We entertain passion, politely, a distant relative in for lunch. *Yes it is, for this time of the year,* we say. *And how are William and the boys?* A short-story writer I admire says, "Winter itself was dangerous."

Yes.

The afternoon sun streams in through the wicker shades on the breakfast-room windows. It is cold to the touch. In the blizzard, the wandering Jew, sitting on a window shelf beside my desk, froze. The roads here are in retirement. And I am in exile.

2.

I stand at the kitchen window as though I were looking into a huge aquarium and something notable, something yet unidentified, might at any moment swim into view in January's unpredictable currents. And I myself am pale underneath. I may float belly-up in this austere month.

3.

The cats are out today, perching on the woodpile, licking their fur as though this were any ordinary month. The migrant cat came in on last January's worst weather, brought in by the Labrador, Duchess, held gently in those large jaws and dropped at my feet as though I'd brought her down over the grainfield, one of a covey of cats. She balloons with fur, has black rings around her gray tail. She's unnamed, lives in the woodhouse, and has a not-unhealthy proletarian bias against the imagined good life led by Puck, the house cat. She shows this periodically by cuffing Puck once or twice then yawning off about her business.

Puck is one of a dwindling number of New Burlington survivors. She was born in a farrowing box on the farm at the village's upper end. Although her mother was feral, Puck has received the legacy of hands. Certain things, once extended, cannot be reclaimed. The simple act of touching is a moral one. And, therefore, wakefulness itself a commitment.

I like these cats. They're keepers of grace in a subtle season. Puck sits among the plants, a piece of statuary. She's like them—made for admiration, not for love. Cats keep their distance, do as they wish, and so we are a nation of dog lovers. It's a time of shortages and inflation. We're not ashamed to buy love.

The unnamed cat floats through January in the ark of the woodhouse, waiting for the snows to recede. In my mind, all the hilltops are dotted with various patient creatures waiting for March. But this is Ohio and we have only suggestions of hills. And we are impatient.

4.

I've turned off the weatherman. I want the truth, not theater. And where's the truth but in extremes? Give me this drama unadulterated.

My neighbor's daughter, aged four, listening to the weatherman, thought he was telling her of "long-range forecats"— an animal, her father explained to her, that was a cross between a lynx and an alley cat, an adaptable creature with the ability to hide out in all kinds of weather without taking welfare.

The weather bureau was a neighborhood invention. The director of the Cincinnati Observatory, Mr. Cleveland Abbe, began giving out forecasts in 1869. He persuaded the chamber of commerce and Western Union of the practical nature of his predictions and, with their backing, started issuing regular forecasts. He called them "probabilities." He said, "I have started that which the country will not willingly let die." A few years later, a congressman proposed abolishing the weather bureau on the grounds that one of his constituents made better predictions with a sourwood stick.

"In nature's infinite book of secrecy a little I can read,"

says the soothsayer in *Antony and Cleopatra*. A modest man, a tone one could believe.

5.

There's no distraction to this month. I wear it, like a poultice. Josephine Johnson says the true, appropriate gift of extreme cold is "the narrowing down." She says, "Withdraw to one room. Put on your reading glasses." Oh, Josephine, but that's for peaceful days, and, as you mentioned, January's a two-faced month—Janus, the porter of heaven, his two faces the gates that go back and forth. Closed in peace, open in war. And so, what's for the unpeaceful days? My kitchen's a stone cell on the unpeaceful days. I pace in front of the fire, turning at the walls. This dungeon of a month when the turnkey, March, is nowhere in sight. And I'm on short rations.

6.

I complain, but I see as far as I wish to these days. At 4 A.M., the thermometer was at thirty below. An hour later I looked again, wishing perversely for it to drop still more. A lashing at the mast of austerity. What can we stand without giving away allied positions?

The world's more manageable after a blizzard. We're down to lard, flour, salt pork, and kerosene. We own what we carry. We're shopping nearby; the world's lost its enticement.

We need this glacial weather. It sobers us up. Florida's too easy, too accessible. No one should live in such a place unless they grow vegetables. A good winter makes us get our backs up. The blizzard brought us up short. The biggest man, the largest equipment. All hunkered down and waited a bit. For a moment, our footprints were erased from the earth.

I could not see the cornfield fences. The barns may or may not have been there. The cat retreated to the basement, slept in the washing machine, waited things out. The chimney shrieked when I opened the door to the stove. Snow blew under shut windows, made large white rooms of woodhouse and barn. The wind-chill factor was fifty below.

Nothing moved within my sight for three days. I could have been a Viking, looking for the edge of the world. We're off the map: *Here be dragons.* I may have to discover this country, my town, all over again. I consider inventions. What would I choose, what discard? I'd not, perhaps, invent paper, and convert us all to barterers and storytellers. I'd bind myself in the leather of my intentions and go door to door with tales in a minor key, and a pitch pipe.

You'd hear my reedy voice through the orchard, and down the long lanes.

7.

I have been reading in the Baker diaries, which I've brought up from the basement assembled in decades, the first part of this century in my hand.

On the twenty-seventh of February, Mr. Baker fixed a ditch and commented on the war in the Turkish Empire. "Trouble seems to be perennial," he wrote. The great flood came in March, washing away railroad bridges, telephone and telegraph lines. There was no communication with anyplace. "We are as isolated," said Mr. Baker, "as though we lived seventy-five years ago."

In April there was a black rain. In May he fought dirt clods in the fields. ("The struggle is titanic.") There was a drought in August, in October a cow choked on a piece of pumpkin, and in November Mr. Baker cursed the city dwellers of Cleveland for their lack of hardiness during an early blizzard. Mr. Baker called them "ornaments on the human race," and said that "the lure of lights and sounds is a disease of the mind."

In December he bade farewell to the year: "It is ended and it is safe to say everyone is glad of it. It was a hoodoo year."

An average day for Mr. Baker is related to the winter day described by H. L. Mencken as one in which "the whole world looks like a Methodist Church at Wednesday night prayer meeting."

I like to read in these crumbling journals the entries of this dour old man, fuming from house to barn and back, his ill humors melting a path through the untenable weather.

8.

Tonight the snowplows are out late, past midnight, nosing through the furiously frozen countryside. I hear that huge bright blade singing into the snow a mile off, on the other side of the S curves. Like the passing of a grand ship, it startles me from my desk. In the morning, there'll be large piles of thrown snow, like the beginning of a wave across the field.

E. B. White says that the plows do a fine job with the highways but they hurl his driveway full of snow. His plan calls for each big plow to be attended by a small plow, "as a big fish is sometimes attended by a small fish." I see it as simply one more of January's sacraments. I only dig out when I need out, which act makes me consider my destination.

Mr. Berry, one of my neighbors, is not as forgiving as I. A bit deaf, he was shoveling out his driveway when the plow rumbled past and threw a wave of snow over him. I drove up as this happened, finding Mr. Berry stooped slightly forward, his cap pulled judiciously onto his head, standing quite still. He reminded me of one of those iron jockeys one sees in this part of the country, those which, in a well-meaning but somehow silly gesture, have had the faces whitewashed. Give a man a big piece of equipment, says Mr. Berry, coming alive, and it brings out the worst in him.

I love to hear the plow out at night, in late hours, making its particular rumbling sound off in the distance, lights blinking through the spray. It pulls me like gravity to the window.

When I was small, I slept upstairs alone, in a farmhouse much like this one. In winter, the branches of a huge oak scraped against my wall at night and I was often afraid because I was so separate from every living thing, my room black and isolated and still, so still I could hear the voices of my imagination outside my mind. No noise but the branches, the downstairs as far away as sanity in a nightmare. And then, at one o'clock, the train came through, blowing its whistle at the crossing a mile away.

The sound restored me. Life going on. Coming through. Make way.

9.

Duchess spends some time each day walking on the high frozen drifts where she assumes various poses. With her ears blowing, she resembles the Ancient Mariner, staring out over a sea not of her own making, January around her neck like an albatross.

10.

The day after the blizzard, the new seed catalogues arrive, as though by design. They sprout in winter-bound mailboxes.

On the cover, a shirt-sleeved man beside a white picket fence pauses in the planting of an apple tree to dream of pie. It's a simple dream, and while not exactly elegant, neither is it inelegant. But I'm a cynic this morning, out from under darker dreams, short of breath, wise to health. I'm in no pieful mood. It's a *middling* dream, neither here nor there.

This is a catalogue of promises: thorn-free blackberries, a practical almond, giant pears on a dwarf tree. There's an all-American cucumber and an apple tree that comes with a pedigree. Fruit topples into pies. Vegetables lie down meekly in canning jars. But I go to the garden with bared teeth and a bandolier. It's a jungle out there.

The pleasure in this catalogue comes from knowing the broccoli is called Cleopatra. "Would I had never seen her!" says Antony. And Enobarbus replies, "O sir, you had then left unseen a wonderful piece of work; which not to have been blessed withal would have discredited your travel." Equally lyrical, my catalogue says of Cleopatra, "Stands up to cold and drought; tight, finely beaded heads; extra early, highly recommended." True, Enobarbus. Age *cannot* wither her, nor custom-canning stale her infinite variety.

What I trust in this catalogue is the photograph of the old judge, the company founder, first in a line of seventeen gentlemen, six generations of nurserymen, each looking more like an accountant, as though out from an experiment in cross-pollination. The judge is a grim, lean-faced old man, a stare off the page that would wither a fruit moth. Talking to *him* over the fence, you'd know where you stood.

Ah, I'm a carper today. Unable to pay, I'm looking for a free-for-all. The catalogue's warm to the touch, full of earnestness. I'd hire poets for copywriters, portrait painters for the vegetables, and bind it all in leather. It's a fair and decent measure against a museum with blue sky in a bottle, and a swatch of clover field under glass.

The world's a place, not a home, wrote Josephine, wisely.

11.

The cold goes on and on. The world cools infinitely. January presses down on us. Newton's apple would have rolled from his hand, to flatten on the ground, proving there's a stern gravity to this month. Eugene Sandow, legendary strongman, could not lift this month from his massive chest.

I warm myself frivolously at words, and outside the mercury is a red line down the wall beneath the thermometer. "Gust and foul flaws to herdsmen and to herds," wrote Shakespeare, sometime well before the invention of central heating. It is below zero and the night just begun. The windows are covered with ice on the inside. They crack as though they'd been struck.

I'm uneasy in such nights. I'm a gruel of emotions in the bowl of my senses. I'm awake and fitful. I listen for sounds of . . . what? Something may break, I think. If it's cold we've never known before, then expect behavior we've never known before. As if radical behavior occurred at certain levels of the mercury. Ellsworth Huntington, a geographer, argued that the intemperate climates—within the temperate zone—produced cultural progress. He wished for a "stimulating" climate, found California's monotonous, pronounced England's fine. Although Charles Kingsley said, " 'Tis the hard grey weather breeds hard English men."

And so we have lust at eighty, revenge at ninety, and above, the blood boils completely. Below, the polar extremes, and what is reserved for the icy bottomings beneath the frost line? Something slow, and old. The other end of what we admit.

The snow squeaks underfoot, and overhead the stars retreat. To teach us perspective. When I was in the seventh grade, Sonny Verdin told Mrs. Hopkins that the world was shaped like a banana. And we would grad-ye-ate and learn to peel it. O seventh grade, when the world's our fruit. And there's an apple for every applicant.

12.

When the mid-January cold comes (I watch for the fourteenth as though it were a day when the bills are due), the house grows taut underfoot. As though beams turn to porcelain and we cannot afford to take a false step. Last night one of my neighbor's hogs froze to her box. And when he moved her a layer of hide stuck to the wooden floor, a poor, thin carpet.

My neighbors are huddled inside at night, artificially warmed by the paltry heat of television. Housewives cluck over the party lines, hatching sociability. And outside, the land seems pale and brittle. Metal rings in such cold. It sets in the bones, a virus. We are not used to such. The air burns. In this air and at these depths, the impure solidifies, settles like a clinker.

My older neighbors lean toward Florida. Like green plants in a window, root-bound in their circumstances, they check before the mirror each morning for thrips, whiteflies, mildew, and leaf mold. What *is* the zip code for the temperate zone? I do understand this instinct to spend the winter in St. Petersburg. It is the instinct that propels the migrating birds, a pressure in the bones. It's a time that binds the limbs, dreaming of heat. The old become like statues. Mr. Lemar, down the road, used to say, "If I make it to May, I'll make it to December."

It's weather that is unspeakably old, primitive. It's *biblical* weather, demanding an arm and a leg, although the Bible says, "Wool like snow," which means to protect. Ah, Ohio, this landlocked place. We're too far north here. Too close to Canada. We're like relatives in the same town, suffering by proximity. It does us no good to hear about the Dakotas when we have twenty-five below.

The moon, full now, climbs the maple. Is there gravity enough to pull us from this month?

13.

A cloud of sparrows rises from the raspberry bushes along the fence line. Where've they been? Wherever, it hasn't been enough, for I've found them frozen this month. In the diaries, Mr. Baker, on a below-zero morning, rescued a cardinal and a dove whose feet had frozen to tree branches.

Birds have tiny but strong muscles in each leg that automatically lock the feet in place. The bird cannot fall, or move, until the brain tells the muscles to move. And so birds sleep, lowering their feathery bodies around the legs and feet. In the human creature, as with the frog, muscle action is often involuntary. We may leap at the slightest provocation. The involuntary muscles have often taken us miles before the brain is up.

As for birds, anything small is a sparrow to me. I've tried to pay attention to them but I haven't the interest. Books on identifying birds are like books on electrical wiring. I have a kinship with my neighbor Mr. Berger, who says, "I don't do electricity." I don't do birds. I put out seed for them, but I rarely watch to see who shows up.

This morning, a tiny sparrow landed on the windowsill, leaning against the pane as though it could feel some of the kitchen's heat, feathers puffed in outrage against the cold. Tiny beggar.

They're all small birds, except yesterday morning a pair of doves came to eat seed I'd thrown on top of the ice. The doves were stately, slow eaters, mannered. They ignored the sparrows, food throwers and tale-tellers at the breakfast table. One dove stood at one end of the trail of seed, the other at the other end, elegant bookends holding up this rabble of sparrows.

When I think of birds I usually think of an old lady I knew who once gave her ailing parakeet an enema with water from Lourdes. And it did not recover. Who *does* tend the dove? Who sends it out?

14.

The front porch thermometer was seventeen below at 4 A.M. The hour of the wolf. Not night, not day. A dead, quiet, air-locked time. Like this month. I've heard it said that 4 A.M. is the hour most deaths occur. I'd think it true of this month, too.

Everything seems fragile in such cold. Wood and metal alike seem to vibrate. True, deep cold. Cold we have only a racial memory for. Ice crystals on the windows shine in the lamplight.

When the months-long winter darkness comes to the Arctic, the natives spend long periods of time sleeping. Those extreme cycles of light and dark cause, among the Eskimo, a syndrome of melancholia—followed by a speaking in tongues and ritual movements of the body—which ends in a generalized seizure. And afterward, a forgetfulness that it happened. Do *we* do things in January we cannot remember?

I, too, want to sleep. Hibernation is the bones' announcement to the body of an untenable position. If this cold persists, I'll be forced to create a warmer universe on paper. Work by barometric pressure.

The cold could be a joke: an old man recalling the worst winter of his life. A scientist says that, cell by cell, the body is redesigned by extreme cold. Will I recognize myself in the spring?

No traffic on the road for three days. The governor proclaims a crisis, a public poverty. But he gives us a picture of a well-fed man demanding leanness, so we are not moved. I want things around me to reflect heat caught in the act: pottery with elegant glazes, an oak table to work on, Vivaldi in a Mahler month.

15.

I understand those old Teutons who burned pyres on the hillsides to prevent the sun from slipping out of the sky. We forget the pain of each winter until it's back again and caught us with our storm windows down. We pull the fireplace up around our knees and dream of spring, an event we seem to recall.

What lesson carries over? It's a hard act to force knowledge to the marrow where it imprints its rigorous instructions. It was winter that made the children of the settlers leave home. The destitute rooms of winter. They left for lights closer together, which seem like warmth even if a lie. Winter allows no fictions. It's a tombstone, bereft of words.

16.

The snow goes on, a trial. We'll learn to endure our paltry selves this month, frozen here in the ice banks of insularity. There's for sure an ardent fascism to January. It's a month of cold cuts, the quintessential frozen glance.

It is possible we may never be heard from again. I am writing this with charcoal, on the canyon walls. We speak of help being imminent but everyone knows better. We sing to keep our spirits up.

I'm going to spend the rest of January in my kitchen chair. It's one of those old stuffed rockers, found in any pre-1960 sitting room. To sink into it is to ask for oblivion. The disappearance of this chair coincided with the rise of drugs.

But I'm suddenly too energetic. I wrap myself in the wool of my decisions, wander down the road on top of the snow. The leaves of the philodendron press against the south windows like faces.

This afternoon, a Caterpillar used by the road crews chugged its way off the road and dead reckoned its way across the fields behind the barn, heading for the next curve over. It sank in and navigated the field slowly, like an old riverboat, its treads kicking out snow like water.

The truth is this: a man needs a flinty woman in January, for the sparks.

17.

The animals have been lively today for the first time this month. The temperature is up to twenty-two degrees. The bones consider possibilities. Six days to Groundhog Day, and I can check the old one that lives under the silo for a five o'clock shadow, thereby getting a reading on the rest of winter. The cat's in heat. There's guitar music coming from the corncrib in the early morning hours. We toughen into February.

Sarah and Dora

"The forsythia is dead," said Dora.

"Who? Who?" asked Sarah.

"The *forsythia.*"

"Oh. . . ."

Sarah complained mildly of her hearing. "You can't see," she said to Dora, "I can't hear. A fine pair. We spend most of the day in the ditch."

Sarah sat in the front window, looking out on the street, watching for the paperboy. The man next door, who worked for the funeral home drove home in the hearse which he parked in front of Sarah's house.

"Get away from here," she said. "I want no part of you. I am not ready yet. I've just bought a new carpet sweeper."

"I went out to a funeral in the flower truck once," said Dora, "and I came back in the hearse."

"Just as long as it isn't the other way round," said Sarah, peering at Dora over her glasses.

When the paper came she read the obituaries first. She always did this. She wanted to know who had gone, who remained.

Mr. Clarence White, of the next village over, had died. Sarah and Dora tried to think if they knew him, or if they knew someone who did know him. "I remember when Adam White died," said Sarah. "He was a miserly man but had himself laid away like the Prince of Wales. We sneaked in and looked at the coffin. There were silk tassels on it. Papa was one of the pallbearers. He wore his chicken-fixings although he didn't like Adam White. Papa said Mr. White was a moral man. He said it cost money to be immoral. . . ."

Sarah had an appetite for the first day in many. For supper, she ate oyster stew and watched Dora cut a piece of custard pie.

"That sure is a God-blessed piece of pie. I never saw a *lady* cut such a piece of pie."

"I made it, I suppose I can eat it."

"You're a *lady*, ain't you?"

"I don't put on style. You eat with *your* fingers."

"I always have. You warm your bottom at the grate, too. I never saw a lady do that, either. Depends upon how cold your bottom is, I guess."

After dinner, Sarah told Dora that she awakened in the night and was afraid.

"Of what?"

"A noise."

"A noise of what?"

"That's *it*. What, exactly!"

Dora snorted. She was a feisty old lady. She cast a cool appraising eye on strangers and looked both ways crossing one-way streets.

Sarah stared out the window. The houses on her street receded into shadows. A child came down the winter sidewalk on a skateboard. "It seems that I have *never* been young," said Sarah, watching. "But I can remember, much to my dismay. If you ask me what I think when I look at my wedding picture, then ask me what I think when I look in the looking glass. 'Who in the Lord's name is *that*?' I say. A woman once said to me, 'You were the prettiest girl I ever saw.' I said, 'You never saw many. . . . ' "

"I feel a draft," said Dora, suddenly shivering.

"That was me," answered Sarah. "I was reminiscing again. . . ."

Mr. Baker and the Weatherman

At night, the weatherman stands in front of a huge, barely recognizable map of the country and bathes us in his passion for fronts and occlusions. He is telling us this is the coldest January on record. Or perhaps it is the snowiest January on record. He struts like an Elizabethan dandy. He's wealthy in statistics. There is about him the strong notion he is giving out weather, like a dispensation.

I don't see for myself anymore, although in times past I have witnessed this nightly melodrama. Myself, I'm in exile. I rarely answer the door. This is a man I wouldn't let in, like someone selling lightning rods or offering to paint the barn cheap. I resent his familiarity, his assumptions. He is selling the weather, and I'm not buying. I want the weather delivered like lines in a foreign film. A shrug of the shoulder, a gesture. I want a drama that filters down, then settles, like January itself. It's a perilous month. I can't be surprised.

For prediction, I go to the cellar, dig out Ralph Baker's diaries. Mr. Baker kept a strict accounting of the weather in the first half of this century. I pick a year, read his January, then look outside. It's as good a system as any.

January 15, 1923: "The wind is strong enough to lift both men and fodder from the frozen ground." Mr. Baker fails to give us wind velocity but we have, instead, a *picture*. Conditions were much the same the following day, when Mr. Baker noted: "The wind is sharp enough to shave with." In January, he husked corn, 2,251 bushels by hand that year, and when he finished, he marked "GOT DONE" in big, underlined letters in his diary. Mr. Baker also sawed wood, hauled some manure, and kept one eye constantly on the weather.

He was unmarried and read Emerson by kerosene lamp in the long winter evenings, and he regarded the weather as a sort of consort, for better or worse. He checked the weather each morning as he poked up the fire, much as a married man might eye his spouse across the breakfast table, to see what *her* weather might be.

When Mr. Emerson referred to the after-effects of a snow-storm as "the frolic architecture of snow," Mr. Baker surely took exception. On January 18, Mr. Baker noted: "Snow squalls unabated, temperature minus five degrees. The stock is suffering severely. A dark, cold, disagreeable, miserable, vile day."

Now if I were forced to choose between those two voices, I'd pick Mr. Baker's stark appraisal. Reading that, one knows that here is a man who has worn January like a poultice. He *knows* his January. Mr. Emerson, by contrast, sounds a bit like a weatherman.

On the other hand, we need the weatherman's largesse of figures, which allows us to spend our way through our various social debts. I see it as a sort of welfare. One does wonder how Mr. Baker managed. Or anyone in times before. How, for instance, did Lewis and Clark blunder their way into the northwest without the five-day outlook?

A journalist I know was asked to do an article on a weatherman's retirement. He went out dutifully on the man's final day at the weather station, listened carefully, and came back to begin: "There is a 98.7 probability that Mr. Harold Purvis will not be at work tomorrow." I admired that sentence. After all, Mr. Purvis had been hedging with *us* for years.

As for Mr. Baker, suffering in the vacuum of prophecy, he finished the winter like it was so much corn to be shucked, and on the last day of it marked again in his diary, in big, underlined letters, "GOT DONE." He'd done it again, by himself, and he didn't owe anybody anything.

Provocative Vegetables

It is always a particular kind of winter day when the seed catalogues begin to arrive. There is snow on the ground and we are sitting around in the kitchen telling stories about the sun as though it were a past event, on the order of the Depression or the Great War. And outside, the catalogues flutter into the mailbox, a covey of newsprint flushed by the season.

I am convinced it is no accident the seed catalogues arrive on such a day. I think the seed companies plan their mailings for such days, spending millions for a suitable weather prediction, and all over the country, just ahead of a new cold front bringing more snow, men in the employ of these companies are stuffing bundles of seed catalogues into the mail.

The next day, completely off guard, we trudge under leaden skies to the mailbox, spring buried in our minds like an old romance and, *mirabile dictu*, we are assaulted by a Beefsteak tomato, leaping off the cover of the catalogue like a salmon headed upstream.

I am never quite prepared for this moment, and my first reaction is to look around a bit guiltily to see if any of the neighbors happen to be passing. The seed companies, after all, are playing upon one of man's older appetites, that of sowing and planting.

It is the desire expressed simply but well by my neighbor, the beekeeper, when he says, as he often does, looking over the bounty of his garden in late July, "What kind of creature would man be without growing a little lettuce?"

This is a sentence with some amount of depth. To my mind, it ranks with a sentence belonging to Mr. Fred Holcombe, who once said to me, "You do not know what an awful feeling it is to go to bed at night knowing there is not a drop of liquor in the house." Mr. Holcombe was right, I did not know, but I understood that his sentence, like the beekeeper's, was a weighty one, and I have always been ready to defer to a man down with a passion of some kind or another.

So on this day, in kitchens all over the land, gardeners sit before their stoves and slip these catalogues out of plain brown wrappers, looking yearningly at pictures of vegetables. What is bad about this is that these are not ordinary vegetables they are looking at. These are provocative vegetables. They are vegetables which may be spoken of as "eager."

In this catalogue, the season is perpetually midsummer. There is a twenty-five-pound muskmelon, referred to as "the Cadillac of melons." It is no foreign import of a melon, puttering along the garden rows on four tiny cylinders, but a Detroit melon, union-built, with a luxurious interior and racing stripes. There is a Lady Godiva pumpkin and a tomato described as "world-beating," a term that bothers me a bit because it implies the tomato has military capacities of some kind, perhaps a warhead.

Squash recline languidly under a bush, and peas burst their pods, ready to leap wantonly into the vegetable steamer. Float-

ing off the pages are things a man dreams of in the deeps of
January and February.

This is country pornography, a book that puts a man in
the short rows. Herein, the planter does not sweat as he plants.
He comes to no grief in his effortless toil. There are no pictures of this man grappling with the mysteries of compost and
cabbage moth. And so we forget all the broken promises of
July and August. We take up spade and hoe and new vows,
innocently ready to go again.

These catalogues tell us (and we want to believe) that if
you can't fool Mother Nature, you can surely tamper with her,
and in most states that isn't even a misdemeanor.

February

Snow yesterday, eight below last night. Dog creeps closer to
the stove. Plants huddle in the south window. I'm feeling *mean*.
If February is the shortest month, why does it go on and on?
We know January for the wretch it is and nod in begrudging
respect to such shamelessness.

But February is a treacherous month. Sitting in the deeps
of February, we could be headed *anywhere*. Everything looks
the same, backward and forward. There's the high light of
February to tease us, yet nothing happens. There's no poetry
for February. *Nothing* rhymes in such a month.

In the yard, the woodpile dwindles, ragged this and that
left over, chunks of an unknown quantity. It is as though the
journey from one season to another were a real move, haul-
ing-men out to load us up, unpacked odds and ends around
the house and yard.

Mrs. Shidaker sits in her parlor window, leans plantlike

toward the pale sun, says, "I'm tired of this snow. I need sunshine . . . and roses. Snow is for the young, not the old. The young can go sliding. I haven't done any sliding for some time now, unless you count backsliding. . . ."

The beekeeper asks his wife, "What kind of month *is* this? Is it the month that comes in like a whatever and leaves like something else?"

"Nope," she says, not inclined to talk about it. February is a no-comment month. She did, however, see a man at the post office who had just returned from Florida. The man said, "I couldn't find no conversation in Florida. Nobody wanted to talk about beans or hogs. So I got homesick, and I come on back home. . . ."

Is this a sign? The beekeeper's wife notices a wasp has hatched in the playroom. There are groundhog paths in the beanfield. A bathing-suit ad in the paper. But the sap-bags on the yard maples flap in the wind. Nothing rises but rumor.

The sugar-makers venture out in a little knot, an erratic body with ten feet. Within an hour, they retreat from the woods in disarray, scattered by the season. They regroup around the woodstove, peer into the soup kettle, have a drink. Mr. McIntire, having misplaced the corkscrew, opens the bottle of wine with a brace and bit. Mr. Spires picks a guitar with near-frozen fingers:

> *Look at them maple-tappers*
> *Ain't they weird*
> *Ice in they underwear*
> *Snow in they beard*

Mr. McIntire comes out, stands in the yard. Blinks. Looks for his shadow. Will there be six more weeks of sugar-making? Mr. McIntire cannot see his shadow. He cannot see the *woodshed*. It is snowing again.

There are reports of cabin fever. People pull up a chair in

front of the seed catalogue and fine-tune color pictures of
squash. A man down the road has confessed to dribbling his
cat.

When the snow stops, I take the old labrador and go into
the woods. I want to see if I can love the place in such a time.
Always learn to love in the most graceless moment possible for
these are the times which rightly test us. I've never trusted the
romance of spring.

Chill in the feet, cold in the head. Perfect symmetry. Feb-
ruary gives nothing, asks for nothing. It's a month of *will*.
Here in the country, we go out wrapped in thermals and the
Protestant Ethic. We bring in March like a new crop, by *work*.

Groundhogs, in Defense of

Over the landscape from time to time, I have noticed small
groups of people talking earnestly among themselves, their
heads bowing so that in this moment, they almost touch. From
long experience, I know what these people are talking about.
They are talking about groundhogs. People around here complain about inflation, but what is really on their minds is
groundhogs.

The groundhog, although sorely set upon, pays no mind
and merely goes on about his business, which is real estate.
This is the source of all the groundhog's problems, because he
has no use for the local zoning ordinances. He will dig anywhere, as if it were his land, and without a permit.

A number of years ago, someone lodged this complaint
against the groundhog: "The groundhog is not only a nui-
sance, but a bore. It burrows beneath the soil, then chuckles

to see a mowing machine, man and all, slump into one of these holes and disappear." This report was not substantiated although my neighbor, the beekeeper, says that he doesn't like groundhogs because he is always losing his children down their burrows. The beekeeper is a man given to odd fits of character, however, and I take his comments with a dose of salts, in the phrase of another of my neighbors who suggests it as the proper way to receive information of dubious property.

This talent as engineer is what gets most farmers' dander up about the groundhog. One fellow told me a groundhog toppled his tool shed. They're in the beekeeper's barn now, and he shoots them and drapes them over the fence, as a warning to the others, he says.

It does seem no one has anything good to say about the groundhog. Somewhere a farmer was quoted as saying that the groundhog "eats to give himself the strength to dig holes, then digs holes to give himself an appetite." As early as 1883, people were being provoked by groundhogs. Something called the New Hampshire Legislative Woodchuck Committee (people around here call the woodchuck a groundhog, but we're all speaking about the same menace to rural stability) pronounced the creature "absolutely destitute of any interesting qualities," and slapped a ten-cent bounty on him. "Its body is thick and squatty," went the report, "and its legs so short that its belly seems almost to touch the ground. This is not a pleasing picture." That is a general picture, though, and it fits any number of aldermen I have seen in my time.

About the only time the groundhog gets any peace is early in February, on Candlemas Day, when people, unarmed for once, take a look at him to see if he leaves a shadow. According to long-established lore, if he sees his shadow he promptly retires and we are left with six more weeks of winter. If he does not see his shadow, he stays out and we have an early spring.

I suppose the reason is that if the day is bright, the weather is likely to be clear and cold, with good prospects for more cold. And if the day is cloudy, with perhaps those dull, low February rain clouds that around here mean good sugar-making weather, then spring might be at hand. This isn't the best reasoning, though, and I recall reading that five days after a shadowless Groundhog Day in Bismarck, N. D., the countryside there was a minus thirty degrees and dropping. This, in fact, may explain some of why people take such a dislike to the little fellow; they are upset about the forestalling of spring and out looking for a scapehog.

The real fact in the case is that the groundhog is a lay-abed, and rarely up by February. A naturalist once told a friend of mine, "Why, on Groundhog Day you'd have to dig one up to see him and even then you could bounce him like a basketball and he wouldn't wake up."

So he is poisoned, shot, trapped, clubbed, stoned, run over, and chased by dogs. I have heard accounts of all of these methods used on groundhogs, and some more. I have never heard of anyone hanging a groundhog but I don't doubt that someone somewhere has tried it.

"Dogs is death on groundhogs," one of my neighbors tells me, the owner of a bulldog, although his favorite way of evacuating the premises of groundhogs was to keep a rifle beside the kitchen table. The table looked out over his cornfield, and he was known to fire off a round or two without leaving his plate. His wife was a placid soul, and I heard their son say they were attending to supper once when the old man grabbed the rifle and fired at a groundhog, aiming across the gravy bowl. "You want your buns toasted, Orly?" this imperturbable woman asked by way of reply, ignoring the echo of gunfire still ringing in her kitchen.

Most people around here simply shoot groundhogs, although most leave their kitchens to do it. One fellow I know hunts them in an old truck. He and several companions will drive around in the fields until they spot a groundhog, then they'll chase it in the truck. The groundhog dives for cover but, being a curious sort, he'll soon stick his head out to see what all the commotion was about. The fellows, meanwhile, will be parked just outside, the motor off, watching. When the groundhog resurfaces, they fire away at him, and in the doorway to his own home.

This seems a pretty fair violation of the unspoken rules between man and beast, although the fellows chose to see it as a form of rodeo. I'm on the side of the beekeeper's wife, who doesn't really have anything much against the groundhog and prefers hand-to-hand combat. Once, when a groundhog walked up unbidden on her porch, she growled at it. She said it just seemed like the thing to do at the time. The beekeeper, who had gone for his gun, was incensed. He'd seen it as another trophy for the fencerow.

Mr. Wills, who lived nearby, once tried to shoot into a groundhog den but ended up in ignominy. The blast didn't bother the groundhog, but it sprayed dirt over Mrs. Wills' freshly hung wash and she chased Mr. Wills out to the barn with a broom. Groundhogs once made a den under the base of my living-room fireplace, and I chased them off by attaching a hose to my car exhaust, dropping it in the hole, running the engine for a half hour, then filling up the hole. I assume that due to the air quality index, they left and are now some of the same family out digging in the barn.

Groundhogs do not have such a reputation, but they are a courageous sort, and fight well and hard if put up against it. I once read a newspaper account of a groundhog coming out on the winning end against a bulldog. One of my neighbors has a terrier of uncertain lineage who spends part of each spring and summer sleeping on the hearth while his face mends from sorties made against nearby groundhog dens, and another neighbor once ran between a mother groundhog and her den and the groundhog promptly bit into his shoe. This man repented forthrightly of his foolish position, and said he should have known better.

Like the beekeeper's wife, I prefer hand-to-hand combat. When they dig around the buildings, I try to persuade them to relocate. My old labrador is a one-man grievance committee, bearing the groundhog perpetual bad news. He also likes to dig, and with his various attentions, most of the groundhogs here prefer the relative sanctity of the woods and fencerows.

On the whole, I find the groundhog a much maligned creature, and victim of a bad press. He makes a good pet, and I've heard of one being nursed by a cat. The groundhog is a clean animal, a vegetarian, and a fine meal if prepared right. He aids in a certain amount of soil improvement, and a host of other animals—even the pheasant—uses old groundhog dens for shelter. I've observed him to be brave, modest, and intelligent. These are qualities I'd settle for in any neighbor and on good days, aspire to them myself.

The beekeeper still argues with me, and his fencerows continue to bear fur. His latest argument is that the groundhogs eat his beans. I recall they ate a quarter-acre of Mr. Thoreau's beans, too. He labeled the groundhog as enemy, but upon reflection he recanted, suggesting that beans grew partly for them, too. "The ear of wheat should not be the only hope of the husbandman," he said, "its kernel is not all that it bears. How, then, can our harvest fail?"

Nolo contendere, beekeeper.

Crime and Punishment, I

A new study reports that crime is increasing in rural areas and although it may not be a pitchfork jungle, the tradition of unlocked doors in the country is coming to an end.
— *Los Angeles Times*

The town is *there*. I am *here*. There are eight miles between. What of this mean distance? Once, a man driving a buggy used an hour of his day going this distance. "Depending, of course, upon the horse," says my elderly neighbor to the south. Today, there is another dependence. A man goes this distance in less than fifteen minutes (it is a country road, guilty with the diversions of curve and hill) but his going depends upon the whimsy of politics.

Along the way are houses he has never seen before. Are allied forces bivouacked behind the unreadable windows of the ranch houses which have sprung up in the cornfields like new weeds? Or is ransom demanded? What are we being asked for safe passage through the intermittent tides of unpredictable seasons? (O Ahab. Today you would not have the luxury of playing your pride-maddened self in the world's greatest fish story. Your brilliant passion would stray to alcohol or neuroses—for these are days like that, Days of Wine and Neuroses—and at best, you would climb, encumbered by your madness, into a bell tower for a private toll. These days, Ahab, you would be what? A charterboat captain? Eating by yourself in a mess of your own?)

To the north is Graveyard Road. Ten years ago it began off the main highway, circled discreetly through farmland and woods for two or three miles, disappeared momentarily in a covered bridge, and ended beside a graveyard. There were eight houses on Graveyard Road ten years ago. Today there are twenty-seven, if one does not count the evolving foundations of the twenty-eighth.

Not long ago on this road, an elderly couple was robbed by a man who came to the door asking directions. He stayed to give some of his own, a modern characteristic. When invited inside, he pulled out a gun and tied up the man and his wife. "No one could have been more surprised," said the lady afterward. "I had simply never thought of such a thing. I was so surprised I did not take the time to have a heart attack . . ."

Now she and her husband check to see that the door is always locked, something they never did before. When she walks down the road to visit her daughter, she locks *that* door behind her, too. She admits that, quite often, she wonders if someone is *watching*.

"There are many alarms in the country," says my neighbor to the south. Although her husband has been dead for ten years, she places his hat on the newel post in the front hall and leaves a light on to illuminate it. She admits that this minor oblation makes her feel more secure. She has a friend who sleeps with a cowbell under her pillow, considering dissonance weapon enough. This is not entirely a whimsical notion since dissonance is a sanctioned contemporary weapon, a growth industry, in fact.

It is reliably reported by naturalists that the sightless bat finds its way by projecting its tiny but purposeful voice which outraces it even in flight to bounce off any object ahead to return and strike the bat with knowledge. The human animal, in like manner, casts his voice about for: direction, affection, and a good under-the-counter bargain. His voice, too, outracing him to bounce off objects like a boomerang. This explains why the human creature is often flattened by an echo. Echoes have both density and velocity and are therefore subject to the laws of physics. Of course, the human voice is now used mainly as a weapon, and is in the process of becoming a refined instrument of some subtlety. There is, for instance, the small-bore contralto, the double-barreled bass, and the .44 magnum opus.

For years, the human voice has been turned up by archaeologists sifting the dust and ashes of drifting centuries. It has been uncovered intact beside the ruins of rack and iron maiden, obscure political manifestoes, and the brass shower-heads of Auschwitz. Certain experts predict that the violent forms of expression are currently in a state of evolution, heading toward a refinement in the human voice. By the turn of the century, it is predicted that the human voice will render obsolete: the fist, the switchblade, martial arts, and the Pentagon.

Already children grow up speaking to one another in the simple parables of advertising craft. Magic is forsaken now for Madison Avenue's intravenous imagery. With clamor as sophisticated weaponry the language becomes fallout, and a virulent residue for unborn generations. Soon we will look backward, longingly, to the clean cartharsis of the thrown fist. Only heretics using the forgotten forms, i.e., poetry and the short story, will warn us of the origins of violence in the crafts (journalism, for instance), the healing practices (psychology, for instance), and even sociability itself.

Those who practice language (steel-eyed, aiming down-wind at a moving target) have brought about a new industry whose stock is rising. Meanwhile, we lock our doors, as a reflex. The latch string is no longer out. In our day we have seen the annulment of the Open Door Policy. We are witness to the end of the neighborhood and the beginning of the enclave. We are becoming a nation of cliff-dwellers, leaving behind for the archaeologists, fear, like an artifact.

I myself find the cowbell an interesting idea, and probably as much weapon as I can safely handle. I'm thinking of con-sulting my hardware man about an all-purpose cowbell for use in classrooms, temples, at cocktail parties, for the proper punctuation of paragraphs, both my own and those addressed to me.

My neighbor reports that she feels truly secure only twice a year. These are times when the farmers come out to plant, and again when they harvest. "There are men about all day, you see," she explains. "*Familiar* men."

Although she was a farm wife, contemporary work eludes her. If the men are familiar (she knew their fathers and grand-fathers), their work is not. "I do not know myself in the coun-try anymore," she says simply. "They do not rotate crops anymore. They plant corn. . . ."

There is no more need for rotation, dear neighbor. The world spins but it does not rotate. It has spells and seizures but not cycles. We are on the verge of overthrowing the seasons (we may seed clouds as well as furrows). The lines of latitude and longitude belong to realtors; beanfields no longer multiply but subdivide. Everything comes to us or we go out, getting. The pupils dilate before the feast. We live in a state of smorgasbord and grow nearsighted, farsightedness an unknown craft, like wagon-making, like simplicity.

Nearby, sugar camps close, farms merge, and farmers' sons go into the factories. One road over, the man who owned our general store has retired although, as he admitted, he was not yet eighty and did not have anything else to do. He recalls mud roads everyplace, the first automobiles. He notices that beside him, men have lawnmowers as powerful as horses, and tractors that cost forty thousand dollars.

The marvelous machinery is everywhere, a population. It can do anything, a sobering fact, because when man is granted his wishes they usually involve power. Machinery thus converts desire into horsepower, a formidable translation. Primitive man must have dreamed of machinery, and so doing, invented the wheel.

Several centuries onward, the process of evolution seems suspect. Mountains may be moved, but not men. Catfish walk in the dust. Spring comes in, pale green and petulant. We wear expectation on our sleeves like chevrons.

The old farmers, their workable holdings down to a garden beside the farmhouse where the world exists in craftsmanlike miniature, admire the elaborate machinery in the nearby fields. But the admiration seems reserved, the admiration of a certain distant respect. The allegiance of the old farmers seems elsewhere.

Some might call this "nostalgia." Yet nostalgia may be only a misused word, grabbed involuntarily by one who feels a nameless pain (this is a time, one must remember, of verbal pesticides, of mists of conversation in which the language browns and falls in wilted clauses). "Nostalgia" may be only the suggestion that our present rhythms are untenable. The body, fragile enough, begins to vibrate.

A lady I know has spent her first thirty-five years conquering a fear of elevators. She has yet to fly, or ride fearlessly on interstate highways. She is holding out for the navigable distance, something that feels *safe*. Where are the *trains*, she asks. Inside the jet plane, engines roar and the thrust pinions us to one of two destinations. The body braces for resolution, an incredible tension, the bones like wires.

I have a neighbor who still judges a good farmer by whether he keeps his fencerows neat. Such a preoccupation dates a man considerably, but my neighbor knows what he knows. Up-to-date farmers, however, have no time for such effects, and my older neighbor's standards go wanting. He has been known to drive for miles to look at a clover field. "It is all big business now," he says, "and the farmer an executive. He rides in an air-conditioned cab high above the ground his father walked on. Why, now he's a man who can *dress* for work. . . ."

My neighbor is talking about how the crow flies. What he suggests, possibly, is the severe distance between our imagined wealth and our real poverty. Hummingbirds cross the ocean, swallows still end up in Capistrano, the buzzard returns rigorously to Hinkley, Ohio, and we fear for our lives crossing the road. And why *do* we cross the road? Because it is there. Even the median strip seems like a vantage point. We are tourists in the country of the last resort. But we would like to be pioneers again. Stuffed with restless genes, we bridle and bolt, looking for panoramas beyond the ax, untouchable horizons, back to bootprint and bear paw for evolution's mad run, like a dream recurring and us waking in a sweat to reach for the light and stop it. But no light, no light. The light at the end of the tunnel is a Sunoco sign, and there's an energy crisis on. Although "energy crisis" is a reference to natural plunder, what we extort from the earth, it is also a personal crisis (*our* energy, *our* resources).

In our neighborhood, where the land is becoming something not country and not city, we have observed burglary, vandalism, robbery, and arson. We are mute witnesses to the end of symmetry, the passing of private ownership of land, and bad manners on the party line. The lines of battle are drawn. So are the participants' faces. The battle no longer goes to the swift but to the brash. To live, someone says, is to maneuver.

We seem to remember a placid time, yet the narrator of *Our Town*, speaking just after the century began, reminds us of locked doors in Grover's Corners. My aunt, a strong but fearful woman, alone in the 1940s, wanted the screen door latched. Afraid that the out-of-doors might leak villainously in, she rose to check the house each midnight. She did not like animals, the darkness, all of nature. A noon worker under the high and mindless glare, she sweated through each summer over the processes of canning, a ritual for expiation rather than nutrition. We had no larceny, by name, in our neighborhood. I myself was the worst child, a petty thief, hunting down imagery. I would have rolled my grandfather for a picture.

It is difficult to remember the indentured nature of those days. We remember, instead, *space*. And while it is proper to recall such a picture, it is now a more interesting time to be alive for at this moment—first time in our curious history—we are free to determine whether we shall live or die. We will do this by our own hands, craft now a necessity.

Here is a definition of romanticism: the discrepancy

between memory and physical sensation. Memory is a bad dog, leaking on the kitchen floor. Memory chews up the slippers, a Sunday hat. Memory is a grandmother-sniffer. It buries old bones deeply and if we know they are in the backyard we never know where. Memory is treacherous because pain is constant, ongoing. Memory has poor capacity for pain. Memory is today, plus time, minus impingement.

In Grover's Corners, there was no crime, but the townspeople decided they wished to lock their doors. My aunt's imagination led her to latch the screen door. Was this a premonition, or did crime evolve in nature as a passion against boundaries?

One must think today to know where one stands, because even the human parts are becoming standardized. Down the road from me, there is a Mercedes parked in the yard of a farmhouse. Farmers wear brightly colored jumpsuits. A neighbor, whom I have never met, has a Doberman. A friend of mine, who may be the most pragmatic of all, has bought a munitions carrier which he parks in his father's barn.

It is possible that what the country needs, in the face of such, is a subsidized manure spreader. The manure spreader, all too unfortunately, is a much maligned invention, yet it never fails to tell us something we often forget: which way the wind is blowing.

Mr. Baker Fusses into the Year

In early March of 1924, Mr. Baker staggered out from under winter and went into town. He noticed there were no signs of life in the wheat field. He decided that the combination of cold and thaw in February had probably killed it. "Deadest I

ever saw at this season," he wrote in his diary in a legibly
Spencerian hand. "Binders won't be used this season."

March was not his favorite month. "Old winter hangs on
and the older it gets, the worse it seems," he wrote. Then there
is a pause on the page, a slight breathing space. "The older I
get, the worse I seem, too," he added. He had a habit of fuss-
ing in and out of the years. At the end of 1923 he wrote, "Last
year was one of discouragement, crop failure, and increasing
taxes. This is a combination to break the stoutest heart. Let us
hope 1924 [and later it would be '25 and '26 and so on] serves
us better. . . ."

He swore at the weather, the markets, and the Democrats.
He noted that trouble seemed perennial, like the day lily or
the hibiscus. He thought the state was headed into bankruptcy
and he fretted over not having all the corn shucked. When he
finished he wrote, "It ennobles a man not to be a slave to a
corn shock."

On the way to town he noticed there was a Prince of Peace
contest at the Sunday school, with the proceeds going to buy
a new background picture for the church. The topic for the
coming Sunday was: Does the Christ Child Have a Chance?
Mr. Baker thought probably not.

In town, hogs were selling for seven and a half cents a
pound. Mr. Baker noted that hog-growers didn't have much
of a chance, either. He bought a horse collar for two dollars
fifty, put two shoes on the mare for a dollar ten, bought four
sacks of cement and a Siberian gut violin string.

That night, he drove over to Mt. Vernon to the lyceum
where he debated against his brother. Resolved, that dancing
is injurious to morals and should be prohibited. He argued the
negative and won.

The next day he turned the boar in with Old Dishface, the
sow. He wrote that it was an ideal winter day. The temperature

was twenty-two degrees. He cut some wood. "The snow covered everything," he wrote. "Every branch, every twig covered to a depth of an inch. The contrast with the blueness of the sky was one of the most beautiful possible to see. The only regret is that such scenes go so quickly and we cannot keep them with us always. . . ."

Two days later, there was a brilliant display of the aurora borealis. Mr. Baker feared a spell of vile weather. That afternoon, the hogs ran off into the swamp.

The next day, however, he heard the first killdee of the season and regarded it warily. "Signs are everywhere in abundance," he wrote, "telling us everything at once. What are we to make of them?"

Then he did what a man with a proper appreciation of life's coquetries might well do: He went out to the barn and hauled manure.

March

The crows are back now. I say to Charles, Do crows migrate? He says, I think they go to town. They sit in a beech top, shuffling about, casting aspersions as we go under. We're a tiny band of pikemen and sappers, mercenaries in the maple woods, up to our knees in March. It's a season requiring stout hearts and calf muscles to match. We speak of knee injuries in tackling the wily maple. Cartilage operations. There's a gallows humor to our crew. We pronounce the Devil's name for courage. We whistle through this graveyard grove of sleeping maples.

Minus five this morning. No one's out. The farmers, like

swallows, have gone south. It's the modern system of rotation: corn, beans, and Florida. Old sugar camps sit abandoned. In one near me, a family of groundhogs live under the evaporator. Who's left around us? No one I can think of. Mr. Peterson, at eighty-six our senior sugar-maker, says basketball's the damnation of it. "Too many ball games to watch," he says. "Soon's a tournament comes on, why, the water runs over. . . ."

It's twelve degrees at eleven o'clock but there's no wind and the sun is out. While Charles taps the bottom half of one maple, a woodpecker taps the top half of another a few yards away. Like any good craftsman, the woodpecker doesn't like to be watched too closely. He stays a few trees away. He was at work before we got here, stays on after. He's non-union, no doubt. A private contractor, on the wing.

Charles has a bit attached to his chain saw but it isn't working today. We're using two hand augers. The woods, except for occasional outbursts by vaudevillian crows, are quiet. Tom misses the noise. He's a friend of Charles, out from the city to help. He's an engineer; machinery comforts him. Myself, I'm secretly pleased when we're back with ax and auger. I understand the basic components of the ax. I can repair an ax unassisted. Outside, I'm on equal footing with engineers. In the shop, I'm an alien with a visitor's permit. I don't know the language. My fingers get heavy. I could never play the subtle sports. I was a lineman, in the percussion section.

Last year, we used both the saw bit and the hand auger. I used the auger. What's your tapping rig run off of? Buck Arledge asked Charles. That one over there, he said, pointing to me, runs off of chili.

We mark this woods by large, mostly rotting old beeches, light gray and ponderous among the hundreds of younger maples. The beech is a miser, hanging on to its leaves as it can through the winter. All grays and browns and white out here. It's a subtle time. It takes a good eye for winter. *Anybody* can appreciate spring.

Charles's daughter, Beth Ann, is out with us. At five, she's a three-year veteran. We're all pros here. We have a real fine team this year and our tappers put on their overalls one leg at a time. Lot of spirit lot of hustle. I carry Beth up into the woods on my back, a forty-five-pound habit. She's the pounder. She carries a hammer and pounds in the spiles. If ever I get on the school board, says Charles, I'd put pounding into the curriculum.

We measure their progress into adulthood by backing them against last year's tapholes. Beth's almost there; Gregg, at nine, is past. Gregg considers the design of the woods this year. Something in his mind's eye pleases him. The woods, which we organize by lines of tubing, has an aesthetic sense that appeals to him. I'd like to go over this woods in an airplane, he says. I could look down and see the tubing running everywhere and it would look like a factory. Gregg has never been in a factory, however. Such a place exists in his imagination. In his naïveté, he implies that a factory is what we will it to be. Gregg has been inordinately high-spirited since we have gone into the woods. At lunchtime, he hugs Judy.

I am *happy.*

Why are you happy.

I don't know.

Yes you do. . . .

Soup fogs the kitchen windows. Gregg's German shepherd sleeps on the picnic table in the snow. Boots steam in front of the fire. Beth puts a pretzel in her mouth, puffs on it. Something she's seen somewhere. I'm smoking, daddy, she says. No wonder you're so short, he says.

The cold lies all around us. March is a beach and the cold's a tide, in for the month. It hurts the lungs. We're unused to such purity. January and February have come and gone, stone cells for months, and now we're out and asked to expand our horizons. Hold on, we say. Where's the signs for spring? March is poker-faced. It's still a full act of the imagination to consider a shirt-sleeves time.

There's a meadow beside the maple woods which reminds me of the story in Boswell's *Life of Johnson* where Samuel Johnson, newly arrived at Oxford and considering lectures, like libraries, to be taken when need arises, attended one then missed several. Said Johnson: "On the sixth, Mr. Jordan asked me why I had not attended. I answered, I had been sliding in Christ Church meadow. And this I said with as much non-chalance as I am now talking to you. I had no notion that I was wrong or irreverent to my tutor." Boswell: "That, Sir, was great fortitude of mind." Johnson: "No, Sir; stark insensibil-ity." And it is, too, a prerequisite for the maple woods. We're a ragged little band, up in arms against the season. We're out to entice. We'll bring spring in, by guile or force.

The snow comes again, and after it, a freezing rain. A bedraggled pheasant walks along the road, its head down. In Columbus, a man struck with a snowball pulls a gun and shoots the thrower. Our Lady of Mercy bingo has been canceled.

The man down the road calls us *sappers*, an interesting word. It is military in origin, meaning a member of an engi-neer unit trained to execute sapping, which is an extension of a trench, especially one dug to some point beneath the enemy's works. And so here we are, a band of insurgents, out to infiltrate the season, balking at the task of calling up the sugar water and our own lie-abed energies. There's a stern gravity out. The body wants to stay at rest. Or at least, in the kitchen. *Mutiny*, Mr. Christian.

Judy, taking a biology class at the college, is asked by her professor if she knows what makes the sap flow in the tree. No, she says. Does your husband know? the professor asks. I'll see, she says. At the next class, the professor says to Judy, What did your husband say made the sap flow? She answers: He said the sugar fairy does it.

An answer only so facetious. No less an authority than Darwin spoke of sap flow as "that most nebulous of subjects." The theories of it range over such arcane possibilities as osmosis, capillarity, root pressure, and electrical attraction. What we know for sure is that we have tapped the woods and the crows are in from town. And we are up to our imaginations into March.

The pale sun teases us. The rooftops steam although the thermometer is still under freezing. The days warm, degree by degree. The nights seem milder now. A sap-run, says John Burroughs, is "the fruit of the equal marriage of sun and frost." In the woods, the south taps begin to run a little. Charles hears the sugar bird in the cornfields. He doesn't know what kind of bird it is because he's never seen it. But each year at syrup-making time, he hears it. But what did it *sound* like, Charles? Oh, same ole bird. A little older. Not quite as loud.

We're sure to run soon. In the sugar house, we wash and clean. There's mice in the holding pan. Cat footprints across the evaporator. Small fellow taking the tour. Dog sleeps under the finishing pan. See that the cat ain't under the evaporator before you fire it up, says Charles to Gregg.

Steam rises slowly in the pans. It's a process that can't be pushed, like spring itself. Last year, when steam began to pour from the sugar house, an unsuspecting passerby stopped in to tell Judy, M'am, I think that little barn out back there is on fire.

I've burned my wrist on the evaporator, a consequence of improper ministrations. Steam burns more thoroughly than flame. Watch me, the evaporator suggests. Pay attention. I'm nervous here. It's *inside* work. Outside, winter passes.

I prowl the corridors beside the evaporator like a captain in uncharted waters. Water runs shallow and lightly in the front pans. It's transformed by heat, and attention.

We're holding an observance, a transubstantiation. Water to commend us in impoverished times. Lift our hearts by the oblique route of our earthly appetites. To run the evaporator is an act of purification between seasons. Upon it we distill both sap and the waters of our own tiresome currents.

May we all rise new and healthy in the face of spring.

The News in the Crevices

The news comes in and comes in, a tide of events both large and small, and we find ourselves up to our suspicions in head-lines and bailing for our lives. It used to be that the news came in, then receded a bit, but there's a failure somewhere in the workings of gravity.

I contend in the face of all this information that what we really know comes from the inaudible, not the insufferable, but that seems to be a minority view and I am not having my day. So I am in the habit of overlooking headlines to wander in the odd corners and crevices of the paper.

The news from the crevices is usually straightforward stuff with no enlightened commentary by informed sources, and we are left to puzzle out the significance of these dispatches in the best manner we can.

I've recently read, for instance, that the police in a small town in Iowa have had a set-to with rowdy geese. An officer, while tagging an illegally parked car, spied a goose peering around the car's front fender. Suddenly, the goose attacked the officer and the officer responded by macing the goose. The goose then dived into a snowbank and sat on its head. There were reports that geese had attacked other passersby in other parts of town. "Everyone is concerned about drugs and burglaries," said the police chief, "but we've got disorderly geese . . ."

In Chicago, chickens invaded the north side. An animal-care officer was called to a construction site where a group of hens had taken over a townhouse unit. A few days later, he got a call that four chickens were chasing dogs down the street, and after that found twenty-one chickens vandalizing a vacant apartment. "These damn chickens are giving us fits," he said, "and we don't know where they're coming from. . . ."

Now while these are perfectly clear reports they contain the ambiguity that is always lurking about under things. What can be the meaning of an uprising of poultry? Can this be a portent, signifying perhaps that the balance between man and fowl has been altered in some dire manner?

I'm afraid it means only that Hollywood producers will hear and make another disaster movie, about Iowa overrun with malcontent geese. After all, Hollywood did it with both frogs and bees. The beekeeper down the road is still angry. He says that the first question anyone ever asks him is, "When do you think the killer bees will get to Chester Township?"

I've also read recently that in a small English town an extractor fan went amok in a foods factory and blasted great showers of mashed potatoes over houses and cars. "Gangs of workmen," read the account, "took a week to wash down the fallout area."

This is certainly food for thought, and if not exactly biblical in scope, then we must make do. While an avalanche of mashed potatoes does not rank with a voice in a burning bush, it surely means *something*. It conjures up visions of the Pentagon looking into the military capabilities of mashed potatoes.

The news in the crevices seems to go on and on, infinitely engrossing, each with its own little mystery. In Canada, I read, a thirteen-year-old boy has been found to be "allergic to civilization." The doctors said the lad was allergic to all chemicals taken from oil, gas, or coal, preservatives in foods, as well as plastics and other substances. I can't say I haven't suspected something like this. I even know a man who breaks out in a rash everytime he gets near a franchise fried chicken place.

I am glad the network commentators stick to the big national issues and leave the news in the crevices to me so that I may fret over it in my own time and manner. I haven't found a pattern to these dispatches, but I'm not worried yet. Such an influential and immediately observable structure as the sun and its planets has always defied a great deal of the thinking applied to it, also. What *are* we to make of this and that? It reminds me of the time Margaret Fuller, the social reformer and Transcendentalist, arose at a public meeting and said, "I accept the universe!" At which Thomas Carlyle replied, "By God, she'd *better!*"

In town one morning last week about 6 A.M., Mrs. Shidaker and Miss Lucas were awakened by a great crash.

"My God!" said Mrs. Shidaker. "What was that?"

"The morning paper through the picture window," replied Miss Lucas.

"Bad news, I expect," said Mrs. Shidaker.

I still puzzle along in the odd corners of the paper, looking for certain clues, but Mrs. Shidaker's comment has remained with me, a fairly succinct statement on what comes in.

The Late, Great New Burlington

All history, granted a wide enough perspective, is merely irony. In the Paleozoic era, New Burlington, Ohio, was largely lime-stone, at the bottom of the sea. Later, it was forest, a great hardwood forest of oak, sweet maple, beech, and hornbeam. Still later, in the last days of New Burlington itself, when the waters of Caesar Creek flooded into the village's only cross-streets one more time, farmer Phil Hartman drove his powerboat across the bottomlands towing his young son on water skis. The boy skied over Bill Conklin's submerged fence posts and up into New Burlington where he landed in ninety-year-old Merle McIntire's peony bed.

Three generations down, five-year-old Gregg McIntire watched both the water and the Hartmans and suggested to great-grandmother Merle that she NOT *move from New Bur-lington as the U.S. Army Corps of Engineers demanded, rather remain and open a bait shop on her back porch which over-looked the Caesar Creek bottomlands. Gregg's mother, think-ing this like him, laughed. Merle McIntire did not. Unsuccessful at willing herself to die, at least for the moment, she soon met the corps' deadline and moved into a trailer park.*

The mollusca (whose shells provided the benevolent organic sediment basic to limestone) had given way to the McIntires, who in unyielding turn, were giving way to the return of the water. And the Hartmans, in an unwitting betrayal of their stricken village, became the first pleasure-seekers of the Caesar Creek Reservoir.

In 1973 Lawrence Mitchner and I vied for the somewhat ambiguous honor of being the last resident of New Burlington. I won, likely because I was dispassionate about it. They took Mr. Mitchner out feet first, which was what he had always predicted, and I survived him by eight months.

Mr. Mitchner, shuttered off against all the world, was not a good neighbor but I missed the companionable lights across the orchard. By this time, the Corps of Engineers had bought all of New Burlington. There wasn't much to New Burlington because at its most it was never more than four hundred people but I felt I was in the midst of a momentous event, for an entire village (no matter its size) was being taken off the face of the earth.

The only other people who believed it was a momentous event were the old widows. They were believers converted by their feelings, unlike me, a believer by the processes of my own peculiar logic. New Burlington had many widows. Up the hill from the cross-streets there were eight of them side by side, a colony, as though spinsterhood were a species. They seemed to know instinctively that this was a final uprooting. Some of them bore this out by their immediate deaths, as though, inarticulate to politics, dying was the only gesture left.

Mrs. McIntire left a note saying she was going to the grave-yard. She started out, then lay down in the snow and died. Mrs. McClure died, too. She lived in the lower part of the village, just above where the streams met. Her house was full of plants and it seemed as though topsoil left by floodwaters nourished the whole place. "Do you think the Resurrection Lily will live transplanted?" she asked me one afternoon before she moved, touching its leaves.

These were two of the village's most articulate statements, although they were not carried in the newspapers. The newspapers had little to say. The local paper, unused to having an opinion about anything, said nothing. The Dayton papers liked the notion of having a playground in their backyard and wrote benign Sunday features about the disappearance of New Burlington. "A footnote to progress," one of them called it. A reporter said that one of the editors in charge of regional news was looking forward to the completion of the new lake because he was a hunter, tired of driving two hours north to find ducks.

New Burlington, meanwhile, disappeared piece by piece. The fire department bought one of the houses for one dollar, then burned it for the practice. A few were moved up the road and put in a field on small treeless lots where the old frame houses looked out of place, in the way farmers always seem in their Sunday suits. The name of the road the houses were moved to was Cemetery Road. If New Burlington had been in a novel such an obvious device would have been dismissed as a pathetic fallacy. Mrs. McIntire's grandson took her house down and stored it.

There were scavengers, too. They took lumber, stole the doorknobs off houses, dug up flowers, raided the orchards, and clipped the ivy off Louie Wills's chimney. Once, when there were half a dozen families still living in the village, one of the residents looked up from Sunday dinner to see several fellows carrying off his fuel oil tank. Every day, men with metal detectors walked over the village, looking for coins, although anyone could have told them New Burlington was a poor place, its wealth both intimate and unappraisable. The printer's wife, in the meantime, went about the village burying bags of nails.

New Burlington was not pretty in its going. It had a protracted, ugly death. Vandals broke the stained glass windows in the church. Debris lay everywhere. A lady who lived near me ran over a piece of tin in the road. "I found I didn't care anymore," she said. "The destruction was like God's own wrath. I wanted the village to remain always. I wanted something that wouldn't change. I wanted it because I don't like the things that will be brought in as replacements. But at that moment I wanted it all gone, all the mess and trash. It is *ugly*, I thought. What did it look like last year? I could not remember. I was suddenly, overwhelmingly sad and still I wanted it all gone, completely. . . ."

There was virtually no one left when I moved into George Lovett's old farmhouse. It was at the top of New Burlington, down a long lane, a two-story brick house built around 1820 and older than the village. I was caretaker for Mr. Lovett's house in the year of its respite. I suppose I was a scavenger, too, even though I chose to see myself as collector of things no one else desired, something like the opossum on his appointed rounds.

I lived on in New Burlington, through its last fall and into the winter. I cut wood from the old orchards and built a floor-to-ceiling bookcase of eight-by-eight barn beams, as though I were going to stay. The only company were the sheep, four horses, a fugitive cat, and a dwindling supply of chickens, unwitting tithe to the foxes, returning now with impunity as the bulldozed foundations of New Burlington grew over with weeds and vines.

There was no television, no telephone, and, therefore, no news. I could have been my own grandfather. I began to imag-

ine how the settlers felt. The farm buildings seemed fastened
to the earth with silence and darkness, fragile clusters on the
end of a long tether of a lane. I understood why Alpheus Harlan's relatives, out in the world, wrote to him in New Burlington and said, "Get away. Go someplace else. Anywhere." I
was as lonesome as the creek bottoms.

I went to funerals, up on the hill, of old farmers I had
talked to through the dying-down autumn afternoons. I hung
back, silent, behind the leaning tombstones. *What was disappearing?* I asked, watching the cedars in that fierce light.

Truth be told, I did not suffer along with New Burlington.
I have often been accused of improper methodology, and I
bow to that. I did puzzle over the lives, as I continuously puzzle over my own. If that may be called tribute, then I paid it.
New Burlington was not a clear tragedy. The language of the
ecologist did not satisfy. It was not enough to say that a village
and its land wealthy in grain and trees and wildlife was being
destroyed because the engineers had not understood the value
of land and the people on it.

That eulogy was sentimental and did not record either the
complexities or the idiosyncrasies of New Burlington. I think
it may be a fair definition of history to say that it is a record of
complexity and idiosyncrasy, over time. I think it also fair to
say that we have a hard time obtaining such a record. New
Burlington, simply (and not simply), had not known enough
of itself. It may be that this is truly a modern problem: whether
knowing less about our own nature feels safer than knowing
more. I would not want this put to a vote today.

And so New Burlington left the face of the earth in exactly
the same manner in which it came. When my time was up, I
left the old farmhouse and found another, out of the lake area.
We moved the animals (there was one chicken left and she

seemed glad to be going) and Mr. Lovett had the house torn down. He made an arrangement with several fellows, giving them the bricks if they would salvage the timber for him. They agreed and went to work.

It was March and a light snow was on the ground. One of the crew had a taste for drink in his off hours, which affected his on hours. So the others in the crew made him remain in New Burlington, unfortunate and dry, the several days it took to dismantle the house. He built himself a crude lean-to, burning pieces of the old house to keep warm through the still-cold nights.

A lean-to was New Burlington's first dwelling, and now its last. I drove by one night and, seeing the small fire and the man huddled in front of it, I thought first of the picture, then of the irony. Soon, there was nothing left of the house, and the spring growth covered the foundation.

The engineers built a dike which curved through part of the old village. The temporary dirt detour ran the traffic over the foundations of Mr. Mitchner's house and the foundations beside it. Nothing was recognizable. A friend of mine drove down and walked up on the dike. "Where's New Burlington?" he asked the engineers. "You're standing on it," one of them said.

The lake itself, twenty-eight hundred acres, was dedicated last year. The governor came, also a senator and a former governor. There was a lot of predictable talk, full of words such as "service." In the area now is a host of younger, louder voices, talking about commerce in its most modern sense. We are, it seems, at the dawn of a new and wealthy day. I've heard it said myself. Most of the people here have forgotten about New Burlington and they would be only mildly surprised if reminded they were fishing over a town.

Crime and Punishment, II

A young man nearby has just been sent away for breaking into a number of homes, most of them near his own. There were so many burglaries for awhile that the culprit would have soon betrayed himself, in the manner of one of the local teenagers a few years back who took some large fireworks and, in a spate of abandon, blew up all the mailboxes along his road. Except his father's, a fact he recognized momentarily, then soberly set off to blow up that one, too, as a way of destroying the evidence.

Mine was one of the houses chosen by this young man, a fact that immediately marked him as an amateur. Of the several things I may reek at any given moment, one of them is never prosperity. In the young man's defense, however, I do recall that old Adam Ellis went around with his pants held up with baling twine while having several thousand dollars buried in a bin of clover seed.

It is a discomfiting event, returning in the evening and finding your house broken into. It provokes a feeling compounded of near equal parts of fear and anger and that gives way, after a look around, of mainly anger which, of course, has no likely object. My first reaction, walking across the yard and seeing, through a broken and raised window, the barn cat sitting contentedly on the breakfast-room table, was to throw a stick of firewood at the labrador, who had just returned from a hunt and merely wanted supper.

The feeling underneath the anger, however, was one of being invaded. What I seemed to have been suffering in the next couple of days, after the sheriff's man had come and gone and I had installed the labrador in the house while I was away, was the breaking and entering of the notions of privacy.

One suffers meekly the various unprivate moments when abroad in the world, that is, the various abrasions in the minor act of carrying out some business or another. But the image of home is one of repair, one's tiny, split-level castle where the drawbridge can always be raised against the rabble. Unfortunately, this is only one more of the fraudulent little romances we have with life.

Under certain circumstances, even the mind cannot hold to its own privacy, since events these days press themselves to an unwarranted attention. And so, secretly, I was pleased to find the young man had been most interested in my old television set, a vintage model of negligible value which operated like arthritis, picking up certain channels only in certain kinds of weather.

I had increasingly begun to recognize its presence, too, as an invasion of privacy, fighting every time it was on somebody or another's definition of crisis, and I disliked its images and did not want them in my head, which, like a schoolboy's pockets, fills itself with enough paraphernalia without television's professional help.

I've an acquaintance who gets angry at the television, once sticking a screwdriver in it, and another time, taking a set into the yard and blowing it into the raspberry patch with his shotgun. The latter occurred, I seem to recall, during one of the president's fireside chats. The next week, however, my acquaintance feels lonely and goes out and buys another one. He doesn't feel lonely because of not having a television, of course, just that it seems to make him feel less lonely at certain

odd times, and he has the good sense to recognize his premise
here as a dubious one.

So the idea here is that one invasion of privacy canceled
out another. I meant to go around and thank the young man,
but the judge sent him off before I got around to it.

It seems that between burglaries and the television, the
peace will forever be destroyed. The outlook for both is bad,
unless you're in television, or a burglar. Our sheriff, a man
who wears a suit to work and could be mistaken for a business
executive, believes crime is not cured, it is merely transplanted, a kind of social surgery where corruption is grated
onto the next county over. (I mention that the sheriff wears a
suit only because two or three elections ago, the sheriff was a
uniformed man who, when the local editor sent over a cub
reporter just out of college, refused to cooperate. The editor
called the sheriff up and asked what was wrong and the sheriff
said, indignantly, "Don't never send me anybody over here
with a *mustache.*")

Our sheriff now seems a pragmatic man, used to both
mustaches as well as larceny, and busy with transplanting. He
knows this isn't a very organic approach to knavery and I think
I detected a reluctant bit of humor at how the culprits seem to
bound back and forth over the countryside. "I have noticed,"
he said, "that if we get on 'em out in the county, then they
come into town and the city police catches it for awhile. Then
the police get on 'em and they come back out to us."

This is the notion of crime as a sort of volleyball game,
but I suppose it has a certain sense to it. My old labrador never
does anything untoward until I leave, then when I return I
find he's been out wallowing in someone's hog lot and dragged
a three-day-old dead possum into the backyard.

After our fit of break-ins, which the sheriff's department

solved fairly quickly, things seem momentarily quiet and last week there was only one break-in out in the county. The run of life, however, seems against the sheriff. "One of the local police departments called us one night," he said, "and said they'd seen something funny. There was this car driving through town with a calf's head sticking out the window. So we run it down and a couple of boys had 'em an old four-door Desoto with the back seat taken out. They had three stolen black angus calves in the back seat. . . ."

I hear there are tractor-theft rings now which operate like the car-theft rings in the cities, which is probably something we have to look forward to, although most of the villainy here still has to do with burglary, such things as the fellow arrested awhile back for drunken driving out in the highway on his riding lawnmower, and the old violations, usually unreported, which people practice upon one another in close places over time.

I've considered putting a small sign out front telling prospective burglars that I'm not particularly a worldly man, poor in goods, and besides, somebody already took the television, which I do not plan to replace. I have only books and as anyone knows, books are nearly worthless. It is very difficult these days to find a good fence for contraband Thomas Hardy.

I've read that in Cuba, housebreaking is a capital offense. I was all for this approach the first week after my break-in, but I've mellowed a bit since. I'd settle for having the ducking stool brought back and set up on the courthouse lawn, and everybody could gather round and get a good look at the culprit. I vacillate on the point of punishment, however, for while horse-whipping might make me feel better, I am uncertain it does much for the horse.

"We're still living on the earth, but in a different manner," one of my older neighbors has confided in me. Well, yes. It is hard to remember that there was never a perfect time. We're constantly out looking for a graven idyll to worship and while we're out, somebody might be crawling through the kitchen window intent on carrying off the appliances. So it seems to me we're exactly where we've always been, which is how to spot the world's larcenies before they come upon us. In the meantime, waiting for the age of enlightenment, I keep a simple house and a bad dog.

Mr. Robinson's Sober Music

When Mr. Robinson began going to church, he stopped playing his guitar in the beer gardens. But when the church members told him he should write only sacred music, he drew the line.

"Most all music," he said, "is sacred music." He even defended the teenagers of the neighborhood and their abominable taste. The members were friendly to him at church, but he sensed something deaf behind the charity.

Then the minister told Mr. Robinson he should not smoke. "Sir," he said to the minister, not letting the remark pass idly by, "if that is all that marks me against heaven, then I figure I have pretty near got the deal whipped." He said this in the same voice as his singing voice, which was mournful and quavering, something like the low wind before a storm.

"Well," the minister said, "it doesn't *look* nice." Then he began to tell Mr. Robinson about how sin occurred in the mind.

Good God, thought Mr. Robinson, *maybe so, but what*

am I to do when I walk the streets of Xenia, wear blinders?
Sometimes he did not know about the church. Sometimes he
felt he had more in common with strangers at the beer garden.

He was a man made humble by a simple but strict life,
therefore he considered what people said. But what he wanted
was to quit paying attention to what *anyone* said. He told his
friend down the road, "I was raised with law and order, don't
think differently. My mother said, 'You get a whipping at the
schoolhouse, you'll get another one here.' It made no difference what it was all about.

"I had to ask permission just to walk downtown and when
I didn't, she would lay the wood to me, don't think she didn't.
She tied me down awful close. If I had brought a girl home
from school and said, 'Mother, I'd like you to meet my friend,'
why, she would have taken the roof off. I told her often, 'But
mother, it says in the Bible that a man should leave his mother
and father and cleave to woman.' She said, 'But, junior, how
could you leave your old mother?' If the weather was bad,
mother would say, 'It's not fit out for you today.' Sometimes I
missed three weeks of school at a time. She was forty-three
when I was born. My father was fifty-four. You might as well
say I've been raised with old people. . . ."

Toward the end of her life, her mind was bad and the
villagers told Mr. Robinson he must send her away. He said
no. When she died, it felt like *his* life had ended. He had
never felt such loneliness before. His brother asked him to
come stay with him, but Mr. Robinson said no. "I have to get
used to this," he said, "so I'll do it now."

He closed off the front room of the house his father had
built three-quarters of a century before and lived alone. That
was when he began to play the guitar and do odd jobs around
the village. He played in the beer gardens but he never drank

over three beers. He said he wanted his music to sound sober. He wrote fifty songs, then burned them because he thought they were too "old-fashioned."

They were simple country songs, most of them about the denial of love. "Love and a twist," he would say, reaching for the strings of his guitar. "That's what they're looking for today. People ask me about writing lyrics and I say, it helps to have a warped mind. . . ."

When spring came, he moved out on the porch, presiding over the neighborhood from an old sofa. He spent most of the spring on the sofa. It was where he wrote what he considered one of his finest stanzas:

> *Bury my money with me*
> *When I pass away*
> *Don't spend it on a tombstone*
> *Don't buy me a bouquet. . . .*

April

When Mr. Eliot, banker and poet, spoke of April's cruelty, he also won himself credentials as a keen naturalist. It's true, we're serf to this month, and thinking revolt. I know that *commercially* April is spring, but it doesn't always act like it.

Recently, a friend wrote to talk over January with me and he ended his letter with this sentence: "For all the hardness of January, I like to stand outside early and listen to the perfect quiet the snow and cold bring, a sense of immanence maybe, as if this were the prior quiet from which every sound proceeds."

That is a fine sentence, certainly not the kind of sentence

found loitering around even a grammatical streetcorner, but I can't do anything like it for April. Besides, January is a month you can get your metaphors into. It *does* have that kind of quiet, described precisely right by Mr. Ackley's sentence. April, on the other hand, is such a noisy month. There is wind and rain, the sound of things loosening. ("Spongy April," said Shakespeare.) People are beginning to be seen again, peering cautiously out from behind the rubble of winter as though not entirely certain the siege is over.

It's the *sense* of spring that puts me on edge, even as I impatiently recognize the principle behind the slow yielding up of things. Spring, like truth, seems hard-pressed to occur instantaneously, and that seems right to me for I've never particularly trusted Road-to-Damascas conversions.

It's only the weary marrow wanting heat and light again, and the evidence grows to suggest we shall have it. The signs, stingy enough in March, abound. My neighbor's young daughter grows splendidly spooky in the face of spring; she's all legs and appetite, sap rising in her as in the maple. Miss Lucas and Mrs. Shidaker have the tomatoes up nicely in their south windows and the old labrador, as for an accounting, drags into the yard everything he buried last autumn.

In the fields, the livestock, too, seems to have thrown off winter. A neighbor tells me he refers to his livestock in early spring as crack-peepers. "I keep 'em up in the barn most of the hard months," he explains, "and they get used to sort of bobbing and weaving, trying to see out of the cracks. Then when I turn 'em out to pasture, they's still bobbing and weaving." A true countryman, he says, can tell how difficult the winter's been by how long the livestock bob and weave.

This afternoon I drove into town slower than usual, watching the dead land rise to its occasion. Mr. Hackney's goat stood on top of a dwindling haystack and leaned leeward, and at the cemetery, which offers perpetual care from an office in a mobile home, the grass seemed almost green. There's irony both explicit and implicit there, reminding us to consider which things may stand and which may not or, perhaps, if spring is here, can winter be far behind?

There are many signs to spring but I contend it's spring when you feel the year shift underfoot, as though it were sliding away and not enough work had been done. But no matter how tenuous, we seem to be seeing spring in and it may be that the winter solstice, just as the old Teutons suspected, reverses itself by the curious gravity of good faith.

An Essay on Taxes, Woodstoves, and the Underlying Frivolity of Systems (Both Human and Mechanical) and Some Thoughts Thereof: Or How Mr. Berger, Lawyer and Woodcutter, Takes on the Government in a Minor but Principled Skirmish

My neighbor, Mr. Berger, and I have been attending to the remnants of our respective woodpiles and complaining about the government. We are aware that both of these are old and inexact sciences but we go right on, as the human creature most often does in the face of even the most comprehensive adversity.

What the government has done this time is dangle in front of us the carrot of a tax credit then, after a time, remove it from our expectant vision. Mr. Berger, being a lawyer as well as a countryman, is using words like "arbitrary" and "capricious" to describe the situation. I like lawyer words myself, so I am seconding Mr. Berger.

This business got under way last year when Congress started bandying about tax credits for people trying to conserve energy in their houses. This was part of the Energy Tax Act, and it was finally signed into law by the president late last fall. The stipulation that most concerned Mr. Berger, and me by proximity since I often have to listen to him, was the one called the "Residential Renewable Energy Source Equipment Credit." I know that this appellation has no rhyme to it and I'd like to call it something else myself but this is a tale of hard facts and hard facts are found rhythmless in many of their varied manifestations.

Mr. Berger first ran across this provision in his office at the law firm where he moonlights when he is not out cutting wood or keeping bees. Since he had just installed a woodburning stove in his house in Clinton County, he put a keen lawyerly eye to the provision. What he found was that a man could get a thirty per cent tax credit for installing solar, wind, or geothermal energy equipment but nothing for heating his house with wood.

Mr. Berger did what a man is wont to do when faced with the inherent friviolities of his government: he got mad. He came by to see me because I had just bought a new woodstove (the old one, after seven years, had become arbitrary and capricious), and we sat down in front of it and I listened to Mr. Berger be mad.

"You know what it is?" he asked. "Wood ain't *sexy*. There ain't no romance to wood. Shoot, *Africans* use wood. Ain't nobody paying no *attention* here. . . ."

The next day, he wrote a letter to the regulation section of the IRS. "The act," wrote Mr. Berger, "cites direct solar energy, geothermal power and wind power. These three not being the sole renewable sources of energy, Congress added 'or any other form of renewable energy which the Secretary (of the Treasury) specifies by regulations. . . .' It is my position that consistent with the purposes of the act, wood is an obvious item for inclusion. . . ."

This is sometimes the way a lawyer talks when he gets mad. He sent this letter to a Mr. Walter Woo in Washington, the man who is writing the regulations. Then he sent copies of this letter to fifteen woodstove manufacturers. Some of the woodstove manufacturers wrote back encouragingly to Mr. Berger. One or two of them tried to sell Mr. Berger a new woodstove.

Some of the manufacturers followed Mr. Berger's advice and wrote to Mr. Woo. One of the notable letters came from a Mr. Ted Bergstrom, who is president of Thermograte Enterprises Inc. in Minnesota. Mr. Bergstrom pointed out to Mr. Woo that wood was *stored* solar energy *and* a renewable resource. He said wood was available to heat a third of the country's homes which, according to his figures, meant nearly seven billion dollars per year that *wouldn't* go to OPEC. He told Mr. Woo he did not think the tax incentive for woodstoves would reduce the treasury because it would be a one-time reduction in revenue and from then on, a great saving.

Last, Mr. Bergstrom rather slyly mentioned to Mr. Woo that after years of outrageously expensive development, nuclear power had progressed to the point where it supplied about as much of the country's energy as wood did.

Of course, it might be said that because Mr. Bergstrom is a manufacturer he is lobbying for his own interests, a man with his own stove to polish, so to speak. Mr. Bergstrom, however, is an engineer and an inventor, and seems to speak off a platform built on the good footing of logic. Logic, though, has never been a necessity of a smoothly running bureaucracy. A good bureaucracy is run, like the remainder of the country, on fossil fuels with all the attendant problems from poor mileage to high emissions.

I had been watching the participants from a good seat in the bleachers, noting serve and return, reams of paper and telephoned words volleyed around the country, and I found myself paying attention out of a kind of perverse regard of the pure mechanics involved. It was the more or less foolish pleasure a man gets from stacking his woodpile in a certain fashion. I went through the statutes myself, laying up that incapacitated language in as neat a stack as I could, then I put up a pile of the woodburners's facts. Logic seemed to belong to the woodburners. But not a tax credit.

So as not to be accused of hiding my bias under a bushel, let me admit to it: I'm a woodstove man myself, and have been for seven years. I would thus be inclined to err on the woodstove side of things. But the purpose of the tax provisions is to reduce the use of imported fuel. The most immediately available renewable energy source is wood, and that is so for a substantial part of the country. Solar, wind, and geothermal power is fine, but it's costly, often complicated, and takes time to install. Mr. Berger paid $550 for his woodstove, installed it in two hours, and reduced the fuel oil he used this winter from fourteen hundred to three hundred gallons.

I've read that in Maine, Vermont, and New Hampshire, eighteen per cent of all households now use wood as main source of heat. Maine woodburners used up 468,000 cords of wood last year, an amount equal to about 700,000 barrels of oil. Senator Thomas McIntyre, a New Hampshire Democrat who wanted woodstoves listed for the tax credit, said wood each year supplies in New Hampshire the equivalent of 1.4 million barrles of home heating oil.

It seems that woodstoves were left off the list because the woodburning folks weren't organized well enough as lobbyists. This appears reasonable to me, exemplifying the American tradition of the lobbyist's pen being sharper than the splitting maul. I was talking to a fellow down the road about this and he said he was thinking of lobbying for the afternoon nap as an energy conserving measure, with an appropriate tax credit.

Out here, we're feeling put upon, I think, subsidizing the development of atomic energy and watching the oil companies maintain their profits, yet not seeing the same kind of incentives for a man and a house. Tempered by a strong anarchistic streak, I waver on the issue, though. On righteous days, I think incentives be damned, let each man worry up his own roof-beams. On other days, looking at the mess with a distant eye, I think incentives, like party favors, should be granted kindly all around. Incentives not being bestowed equally in nature, let the government come to God's aid. There's an intended twinkle to that sentence, however, that may not show on the page.

Mr. Berger, meanwhile, has been back on the telephone with the beleagured Mr. Woo in Washington, he has written his congressman (who is unsympathetic), and talked to the Wood Energy Institute in Maine. The president of the institute, Mr. Andrew Shapiro, says there is an interesting legal issue of whether or not the secretary of treasury is bound by the Energy Tax Act to expand the list of things that are eligible.

Mr. Berger thinks the legal issue is clear. "Congress made the statutes," he says, "and the final part of the law-making process is writing the administrative regulations, which clarify and refine the statutes. Because this was a very technical issue, Congress said, 'or any other form of renewable energy the Secretary specifies by regulations. . . .' What the heck does this word 'other' mean? They can't just re-state the cockamamie statutes. They have to make *sense*. The Secretary has been told by Congress to specify other forms, and he has to do this in a rational matter. 'Other' means *woodstoves*. In legal terms, to leave out woodstoves is an abuse of discretion. I call it 'crazy.' "

We were sitting in Mr. Berger's kitchen at the time, and he was getting excited and waving his arms around. Suddenly he stopped and looked at his arm. "What is this?" he said. "Here I am, arguing before the high court and I just put my elbow in chicken bones. My God." This is, of course, the reason why I like to listen to Mr. Berger arbitrate: you never know what will happen next.

The following day, I saw in the *Wall Street Journal* where the secretary of treasury had been urged by taxpayers and at least forty-one members of Congress to make woodstoves eligible for the tax credit. But as Mr. Shapiro says, "To date, the Treasury has taken an extremely hostile attitude towards such an expansion. . . ." Tax time settles in like an uninvited relative.

Mr. Berger goes on, writing letters, making telephone calls, and staying mad. "I been mad for four months," he told me the other night. He'd just been down and measured what was left in his fuel oil tank. "Down about *this* much," he said, indicating a miniscule amount. Most of a cord of wood, his fourth, was left in the barn, too. I've never seen an angry man sound quite so pleased.

Syruping, to No One's Liking

The syrup season was not to anybody's liking. First, it was too cold. Then it was too warm. The beekeeper's sugar water was as sweet as ever but there wasn't much of it. A couple of weeks after he hung his buckets, a sparrow was trying to build her nest in one.

The Squire, who lived across the road from the beekeeper, said he knew why the beekeeper got such sweet water. "First run is normally *dog* water," he said. "Dogs water down everything in sight, so that's what you get in the first run. Dog water. Trees got to clear out the dog water first. Then they get down to the pure, sweet stuff. Where we live here, them tree's the sweetest trees in Chester Township because neither me nor the beekeeper has any dogs. . . ."

No one knew whether the Squire believed this or whether he was just passing along a theory which came to him while plowing. This sometimes happened to The Squire. He came up with an elaborate theory while plowing but by the time he got back to the house, he had forgotten part of it.

While The Squire was talking about dog water, he had a slightly bemused look on his face. It was the kind of look he had the morning of the last blizzard when some of his neighbors roared into his yard on snowmobiles. The neighbors were wearing their pajamas under heavy parkas and waving a bag of sweet rolls, demanding The Squire and his wife get up and fix breakfast.

Nothing happened at the sugar house for several days, then a thaw came and the sugar-makers worked for two days without stopping. Someone might suddenly become aware he was running the evaporator by himself in the middle of the night while everyone else had sneaked off to bed.

"I see you got up about two hundred gallons before I did," said the sugar-maker, who had just arisen, to his neighbor who had not been to bed.

Other neighbors visited after supper. They talked about the high price of Scandinavian woodstoves. "Why them stoves so expensive?" one asked. "They didn't *use* to cost that much," said the sugar-maker, "but now there's a wiry little Scandinavian fellow comes with the stove and chops all your firewood." The sugar-maker paused immaculately. "And," he continued, "he's *union*."

The beekeeper came in shortly afterward. He hauled his sugar-water over in milk cans and a huge pickle jar. The doctor, who changed his office hours during syrup season, brought in his sugar-water in a new garbage can. The first time the doctor tapped, he hauled the sugar-water down the hill in buckets, through briars and weeds, and brought it in full of cockleburs.

The sugar-maker laughed, and strained the doctor's water before he boiled it. Since, the doctor graduated to a line running from his grove, down the hill, and into the garbage can, which sat in the back of his Jeep.

After a week of warm weather, the temperature went back to twenty. There was snow. The pump froze in the woods.

The sugar-maker worked on it. "———— winter!" he yelled, at no known object. The diaryman, driving past on the lane, offered a sympathetic grin.

In the woods, an old beech fell, and dislodged an owl nesting in its hollow trunk. The owl reminded the sugar-maker's young son of a story. "There was this lady," said the son. "She lived by herself and didn't like it. So she said to herself, 'I want a man.' There's this owl in a tree outside her window. 'Whoooo! Whoooo!' it went and this lady said, 'Anybody, by God!' "

"Where'd you hear *that?*" asked the sugar-maker's wife, laughing.

"Maw-maw told me," he said.

When the thaw came again, the sugar-maker's wife got the graveyard shift. The windows of the sugar house sweated, and the pattern looked like an impressionist painting. A storm came at 2 A.M., with lightning and rain.

Only the sugar-maker's wife and a neighbor were still up. "I feel like the wind has blown away everything in the world except this damned sugar house. Everything in the world is gone but us. We are all that is left clinging to this cinderball, and we are still boiling syrup. . . ."

They finished sometime before dawn. The neighbor went home, poked the fire and contemplated, with Frost, the final days of winter:

> *Come with rain, O wind Southwester!*
> *Bring the singer, bring the nester;*
> *Give the buried flower a dream. . . .*

A Dearth of Porches

My hands sweat whenever I get a letter from the IRS. I can merely feel inside the mailbox and the texture of officialdom travels through the nerve endings to tell the brain how to respond, which is with stringent misgiving. This is not the proper response but it is my response and I suspect I share it with a great company.

The problem here is not even one of guilt. It is purely an involuntary sweat reaction. Mrs. Lord, a lady who helps people repair their tax returns, has noticed this phenomenon, too. She says that if the taxpayer makes a mistake, the government just sends along a bill. "It is like a plumbing bill," contends Mrs. Lord. "People don't sweat when they get a bill from the plumber." I understand what Mrs. Lord is saying. The plumber made a swath through the neighborhood recently and when the bills arrived, no one broke into a sweat. The fellow below me wept, and the one up the road swore.

I like Mrs. Lord's image of government-as-plumber, however. She has given me a picture of a fellow in an old panel truck, setting out to work on the economy with pipe wrench and Roto-rooter.

What I think has happened is that the IRS has been using stationery left over from the draft board. These envelopes, permeated with the compounded power of homesickness, powdered eggs, and infantry life, could account for part of this mailbox *angst*.

We are speaking here of a kind of generalized trauma, which has become an important function of the government.

I suspect this is not mere whimsy and that the government considers trauma part of its normal functioning, providing jobs for psychologists, counselors, and the drug companies. I do not have any statistics on hand but I suspect trauma is a growth industry and, as such, is entirely within the American frame of reference.

For those of us yet to pay homage to trauma as a commodity, I see two immediate ways for the government to spare us. One, the people in the IRS should buy their stationery individually and, when the odd occasion arises, write the taxpayer a handwritten letter. Two, the IRS needs a porch.

I have long been convinced that any of mankind's problems could be properly brought to bay on a good porch. I even spent an hour the other night scanning the papers for facts about where the president and Mr. Brezhnev were going to sign their arms limitation treaty. When I read the treaty would be signed in Vienna at the Hapsburg dynasty's Imperial Palace, I spent another hour trying to remember if such a place had any porches.

The IRS's porch could be either a back or a front porch, but it should have railings, so everybody could put their feet up. The feet need to be up during a good problem-solving, as this takes away purchase for bellicosity.

I think taxpayers should be able to pay in services, also. This is not a new idea, but rather an old idea's revival, going back to when people in the neighborhood paid their taxes by hauling gravel for their roads. I would like to be able to pay at least part of my taxes in firewood, and I think Mr. Craig over the way should be able to pay part of his in goose eggs. What kind of government is it that can't use a few cords of seasoned hickory or some goose eggs?

There is one final thing, too. When I fork over the IRS's

due bill, I would like to do this to a familiar face, preferably someone in the neighborhood, with a porch. That way, if I have a complaint, I'll know where to go and put my feet up.

On Hogs

A man named William Hedgepeth has just written a book called *The Hog Book*. Since I've had a lifelong acquaintance-ship with hogs, I've been going through Mr. Hedgepeth's book to see if he has got his facts straight. My conclusion, now, is that Mr. Hedgepeth knows his hogs.

The thesis of this book is that hogs are clear-headed, per-spicacious beings with feelings. Mr. Hedgepeth refutes the charges continually levied against the hog that he is greedy, lazy, dirty, and pig-headed. Mr. Hedgepeth cites examples of hog thrift and intellect, points out that the hog sleeps less than the average dog, and even takes exception to the phrase "you can't make a silk purse out of a sow's ear."

Mr. Hedgepeth not only dislikes the innuendo behind this phrase, but he says it isn't true. He says you *can* make a silk purse out of a sow's ear and reports that only a few years ago, Mrs. Tinie Smith of Atlanta, after some experimentation, made a silk purse entirely from the ears of sows. He also reported that Mrs. Smith's purse was "quite smart." He said she has aspirations of manufacturing them commercially and plans to send one to the president's wife.

Mr. Hedgepeth contends throughout his manuscript that the hog is *deep*. Good hogmen, he says, are properly respectful of the intellect of a hog. He cites examples of hogs trained to point and retrieve birds and tells us that at least one Roman

emperor trained boars to pull his chariot. "They's just no *tellin'*
what all a pig can do," he quotes one hogman as saying.

Mr. Hedgepeth also delves into the history of the hog, giving the reader a sturdy picture of the hog in fact and legend.
The hog, we are told, was domesticated in 5,000 B.C. by the
Chinese, who are still the world's number-one hog-raising
country although my neighbors in Clinton County probably
wish to disagree with this. One fellow down the road, when I
dropped this fact on him, disagreed vehemently. "Ain't no
Commonist can rightly be a good hog-raiser," he said, shaking
his head in disbelief.

The Chinese, though, knew they were on to something,
according to Mr. Hedgepeth, who says they sometimes buried
whole pigsties in their tombs, the notion being that one should
not be denied hog wealth and comfort in the next world. It is
nice to contemplate the oriental view of things which cannot
imagine a heaven without hogs. Methodists never think this
way and have an awfully sanitized view of heaven, which is
why I've never particularly wanted to go wherever it is they
think they are going.

Mr. Hedgepeth says that British missionaries introduced
hogs to the Melanesian Islands where the natives raised them
fondly, using them both as part of their religious rituals and as
currency. They were cannibals but they loved pork, too, and
thereafter called human meat "long pig" because, Mr. Hedgepeth says, "of its profound indistinguishibility in taste and
texture from the flesh of the shorter pigs. . . ."

Facts like these interest Mr. Hedgepeth, leading him to
assert that hog and man are, as he puts it, "eternally cojoined."
He contends that man is even a bit uneasy in the presence of
a hog for the hog does not subscribe to any of this man-over-
beast business. "There's a lot of things to be learned about
hogs," he quotes a hog student as saying. "There's not too

much good research goin' on hogs today so what we're all doin' is just sorta flammin' around out there in the dark."

Mr. Hedgepeth also thinks hogs do something he calls "smoothing out the raspier edges of society." Iowa is his example, although he thinks his thesis encompasses the whole of the midwest. Iowa, however, a place Mr. Hedgepeth describes as one where citizens brush their teeth eighteen times a day and read the Bible in the bathtub, has the greatest number of hogs. Indeed, there is a full litter—eight being an average litter—for every single person. And Mr. Hedgepeth thinks there is a direct correlation between the number of hogs in Iowa and its absence of antisocial behavior. He points out there are very few riots in Iowa, and he thinks the benevolent presence of hogs contributes to this state of affairs.

"Perhaps," Mr. Hedgepeth sums up eloquently, "the hog mirrors the pathos of the country itself: huge, heroic, maladroit and always straining toward some elusive dream beyond yet another clod of dirt. Thus, both for now and for the foreseeable time to come, there is hope. For there is beauty in the beasts: there is True Art inside them."

No man who ever contemplates hogs should be without a bedside copy of Mr. Hedgepeth's book.

Village Life, Hold the Consecration

I've an eye for the minor ironies. Friends say this might be fatal. They are doing advanced work in the major ironies. No matter, I say. Enough for all. I myself live in three rooms, work small. I hear there are large topics in the world but this could be hearsay.

Within recent memory, I lived in the smallest of villages,

too. Then it was removed so a lake could be built over where it had been, so other villages downstream could be saved. This is a minor irony but it borders on a major one.

Two streams came together underneath my village like a tuning fork, the village itself a single well-tuned note between them. Sometimes I believe this. What the village had, I think, was a kind *form*. My village also had a chicken thief, several alcoholics, a miser, one or two confirmed gossips, and a number of rather earnest Methodists. These visitations upon the human form may be considered *not* kind, but this is so largely for those in the possession. The rest of us could usually escape because the form of the village allowed one to avoid surprises. In the world today, this may be a large virtue.

The village form has been around quite awhile. It is still around but it has been subverted. The shopping mall is modeled after the village. Communes were, too, but they were too romantic to succeed. The shopping mall, as perfectly realistic as a military base, will not fail. And, for that matter, neither will the military base, which is modeled after another kind of village, the feudal estate. Senior Citizen groups and the Rotary Club are village forms. So is vegetarianism and Transcendental Meditation.

The village itself failed not because it was romantic but because we became ashamed of it. We wanted everyone to think we grew up in Boston. As soon as we were able, we went there so we could write home and our parents could show the postmark to the neighbors. Americans have always been like this. Americans are people who either want to move, or they stay in Mattoon, Illinois, and feel defensive. The Pilgrims were at Plymouth Rock four days, then they began talking about how to get to Denver.

Sometimes I think the size of the country did this to us.

We were a people courted by sheer size, like a small man wooed by a large, handsome, wealthy woman. It went to our heads. We incorporated size into ourselves as a virtue. Regard, for a moment, size. Who can, for instance, name the bantam-weight champion of the world? Consider Texas, which we have been taught is not a place at all but a certain way of carrying one's self.

Americans have always been overreachers. It has produced the best of our technology and the worst of ourselves. It answered our questions about getting on, but none about our interior lives. The interior life is almost nonexistent now. It is the attic voice telling us our socks don't match and to watch out for fried foods.

That is why we recently went through a phase of revering Harry Truman. It's nothing much to do with *him*, of course, but rather it's a notion about something called "plain talk." We admire this as nostalgia, an awful fact, because it means we aren't hearing any of it, and don't expect to. But we still admire the *notion* of it. This is the notion of the accurate human voice as endangered species.

We still admire the notion of the village, too. But most of us live in Columbus. There is now a whole literature of the village life. It exists in publications like *Mother Earth News*, which is a periodical for people who feel up the creek without a parable. I take the *Mother Earth News*. And sometimes, when a new issue comes, I'm aware of being titillated. There's some-thing a little prurient about it. It even has a centerfold. You can unfold it and learn all about sprouting.

The first white man in my village was a military surveyor named Anderson. The first settlers were farmers. The village was a union of the sword and the plowshare beaten together. This produced the militant plowshare. In time, it resulted in

the technology of orderly fields. Farmers declared they couldn't
make their way on anything less than one thousand acres, and
to manage the new urban population, we invented the ghetto.

Motion in the American life seemed, for a time, to satisfy.
Now that we've been everywhere and done everything, we're
beginning to think otherwise. People are striking out to find
something called "community" as though the American Automobile Association had the way marked on its lyrical maps.
I'm skeptical of such pilgrimages. They have a way of endow-
ing motion and forgetting destination.

When my village was being demolished, I walked through
the disappearing houses. They were in layers, as though
revealed for some cultural geologist. There was wallpaper upon
wallpaper and pastel colors upon that, and carpet over linoleum on top of hardwood floors unseen for half a century.
That was motion, too. The villagers were being *taught*.

We saw the village as being a restricted place and we grew
ashamed. We were taught that, too. But all the time we carried the restrictions inside ourselves. We turned outward
because the view was easier. On a clear day you could see, if
not forever, then at least past the going percentage.

Americans never learned to make themselves enough real
monuments. My village's first settler was a man named Jen-
kins. When he died, he left the village a graveyard. I've since
forgiven him because he was only there a few years and, no
doubt, felt pressed. Our sense of monuments has always been
curious. Finally we chose large money as a monument, and
that notion cultivated our smallest instincts. And that's a major
irony.

So we're looking at the village life as if it were consecrated,
and that was our original mistake. We've always been too free
with consecration, laying it about in every public place as

though it were a universal currency with the power of purchase upon a moment's notice. I believe in village life simply because one must choose to be somewhere. And I choose, finally, against size: In the narrow life I can watch my flank.

Desire amidst the Clover

What he wanted to do when warm weather finally broke was sit on the porch and watch the clover grow. Maybe he'd plant some, he thought. Then he remembered, he couldn't. He'd rented the fields to that Hopkins kid and the Hopkins kid wouldn't plant anything but corn. The Hopkins kid thought the Garden of Eden was originally a cornfield and the snake tempted Eve with a roasting ear.

What the big boys did was plow up the fields in late fall and let them lie fallow until spring. The freeze and thaw made the ground more pliable. He thought that if the big boys planted something other than corn once in awhile, the ground would stay pliable. To him, without winter wheat, without even the stubble, the fields looked exposed. They looked raw. They depressed him.

The big boys. That was what he called the large-acreage farmers. What they wanted, he said, was six rows of corn from Ohio to Florida so they could harvest going south in the fall and plant coming home in the spring.

But what he wanted was clover fields. One clover field. He wanted the green of it. He wanted to smell it. He wanted to walk in it. Last spring, he had to drive seven miles just to find a creekbottom pasture filled with bluegrass. There was a gate, so he drove his old Buick out in the middle of the pasture and just sat there until some busybody stopped and asked him if he'd had an accident.

"No, thanks," he said. But then he didn't want to sit in the bluegrass any longer, either. The whole world's an accident, he thought driving home, but keeping an eye out for a clover field somewhere just in case the whole world wasn't an accident. He didn't find one, though.

Feeling the winter weaken, he even thought fondly of baling clover. He'd like to feel the heat of the hay mow again. One of the ten steps on the road to senility, he thought, is when you look forward to working the mow. But he could smell the drying clover, and the heat would make him feel loose. Maybe he could lace his own boots again.

Two summers ago, he'd driven the hay wagon for old man Eskew. He had a great time. There was a late afternoon breeze, and the old man's hat blew off and he baled it. Afterward, he stayed around and ate four pieces of Mrs. Eskew's fried chicken. He hadn't eaten chicken like that since. He wondered if there was a connection between disappearing clover fields and the demise of properly fried chicken. He thought there was.

Before he reached home, he decided he'd do something about the lack of clover fields in the neighborhood. He drove into town and stopped by the feed mill and bought several pounds of clover seed. As soon as the frost left, he would plow up the backyard and sow clover.

Then when it came up, he wouldn't cut it at all. He would take his rocker out and sit right in the middle of the yard and smell the clover. Frances would raise hell but that was okay. Maybe he'd put her rocker out there, too. Besides, she still had the front yard.

That night, he placed the sack of clover seed on the bedside table so he could smell it. When Frances asked him what was in the sack, he said casually, "Oh, just some seed. Thought we'd do a little landscaping out back when the frost lifts. . . ."

The Confounding Manifestations of Camping

The campers come out this time of the year, the same as the asparagus, and they're found mostly in the same places, wild along the fencerows. Although I have spoken with one or two of camping's adherents, try as I might, I seem unable to make much of their confounding manifestations.

As I understand it, camping is a way for man to alter the hurried rhythms of his existence into waltz time. He does this by following a metaphorical set of Arthur Murray footsteps back to nature. In the modern instance, these footsteps are things such as restored villages of pioneers and recreational complexes.

It has occurred to me that the perfect experience might be a pioneer recreational complex, an amalgam of the best of the old and the new, such as, let us say, a stagecoach on skis, a disco dulcimer, and perhaps a topless flax spinner or two.

This is, at any rate, the healing movement from chaos to campfire, and while it resembles the notion of evolutionary progress implied in the movement from cave to fallout shelter, it seems to work for a sizable portion of the populace. The campfire has supplanted the backyard barbeque grill as symbol of the new tranquility, which campers go after with a vengeance. As an example, two fellows driving big mobile vans engaged recently in fisticuffs at a campground nearby. They were after the same parking space because it favored the television reception.

On any given summer weekend, there are large herds of these sleek beasts hooting and jostling for position at various recreational watering holes around my part of the country. Some of them even have scenes painted on their sides as if the driver, plunging off into the underbrush and not liking what he found there naturally, could sit down and stare at a sunset of his own devising.

This is an interesting notion, although I've never had much quarrel with sunsets except symbolically, and in that manner prefer sunrises. I did once know a man in the mountains who built his home so that his living room had a magnificent view of sunsets. Then he set his television in the picture window so that it seemed the sun was always going down behind the evening news.

As a boy, I camped out some, mostly simple safaris into a pasture within parental earshot. Toby Hunter and I spent part of one summer camping in a pasture, en route to flunking astronomy merit badge. We could never locate Sagittarius the Archer so we reworked the heavens into neighborhood personalities. Among others, there was Kellerman the Butcher, Hance the Insurance Salesman, and Gooch the Goat Man, all constellations that constantly rotated through our field of vision.

A few years later, after my Uncle W. W. returned home from service, we went camping in the Carolina mountains. Off the beaten path, we found an old abandoned amphitheater and unrolled our sleeping bags out in the audience. Sometime after midnight, I awakened to see by firelight Uncle W. W. standing on the decrepit stage, holding in one hand the bones of a long-departed possum, while delivering what he remembered of the gravedigger's scene from *Hamlet*. It was a fine scene, with staying power.

Years after that, I took a week's sojourn up the Great Miami River, camping out on sand bars and in abandoned riverside shacks. One of the finest night's sleeps in memory occurred on that trip. At the pure deadend of a long day of canoeing, we ran into a storm. As the first of the rain came, we rounded a bend and spotted an old house. It had a roof, a stone fireplace, a kitchen full of dry driftwood, and an old bedspring to put our sleeping bags on. We sank into oblivion by firelight, under rain on the roof.

That's what I remember of camping. I've heard, however, that there is now a camping van with a sunken bathtub and a baby grand piano. I don't think a lot about these new manifestations, but from time to time I worry a bit: I understand there's a Winnebago in my family, on my mother's side.

Small Talk

This is an Old Poot piece. Consider that warning enough and get out now before I ruin your breakfast. They've just dedicated the big new lake down the road from me and I don't care much about it. I didn't go to the dedication but I could hear, all the way to my front porch, the clichés of progress falling like timber. Get two or three officials together and dance to the sounding brass.

I sat on the porch, watching the passersby head for the water like lemmings in their season, and toasted them all. I toasted size and progress and big money and big talk, just as though I'd been there. An irony deficiency is one deficiency I've never suffered from.

I'm for progress but not as much as I once was. I'd like to see something that would cure cancer, or even television, and I'd call that progress. The difference here, however, is the difference between power and will power, a difference that is considerable. Cancer is a disease of aberrant power, which is why people use it as a metaphor for government, and television is a disease of aberrant choice.

If you happened to have been a late citizen of the lake basin, you'd likely say the lake was an example of aberrant power. If not, you're probably more concerned with a boat and horsepower and horsepower is aberrant or benign, depending on its use. A neighbor of mine refers to the new lake as "the motorboat preserve." He finds it intriguing that the government took it away from the Indians and gave it to the farmers, then decided to take it away from the farmers. He says that shows the government is a bunch of Indian-givers and God only knows what they'll do next.

I find myself thinking of smaller amounts of everything, including horsepower. Once I knew an old fellow in the mountains of North Carolina. It was some of the loveliest country on earth, and some of the poorest people lived there. "Me and my boys have everything we wants," he told me. Then he paused a moment. "'Course," he added with a sly grin, "we keep our wants *simple.'*

Progress, to my mind, always means we're about to pave over something and put up one of those little buildings that sells cellulose chicken. "The civilized man has built a coach," said Emerson, "but has lost the use of his feet." Emerson did not, however, foresee the time that Evinrude would build a coach and still let a man use his feet, while being dragged along behind.

I do not really begrudge water-skiers anything, but it seems we're putting water on part of the countryside and trying to pave the rest. Soon we'll be growing tomatoes on the garage roof. I know one doesn't satisfactorily water-ski in a beanfield, but I'm partial to beanfields anyway. Given the whimsy in our national intent, we may one day be faced with deciding whether we want to eat or to water-ski.

What is wrong with a small deal, the small picture, or some small talk? We've a real problem with size. We even seem to prefer, as a male nation, large chests on the women-folk. A woman from France, noting this odd application of ours, told me she thought American men were improperly weaned.

I've thought some on this, and it could be the basis for our daft preoccupation with size. It would seem to adequately explain why someone felt pressed to build the World Trade Center, for instance. Another neighbor of mine had recent occasion to fly in the Boeing 747 and said he could have gotten a thousand bales of good lespedeza in the tourist section. Now that's a definition of size.

I thought some more on progress and size, then I went into town and built a trellis for a lady's vining strawberry plants. I spent a whole lot more time than I had to, using screws instead of nails, and afterward she offered me some fried chicken.

Things were suddenly back to scale but I have no idea how long I can keep them there.

Mortal Occasions

The crowd in the cemetery on Memorial Day was small, a mere fraction of the one underfoot, and not many there remember when it was an *occasion*. A few years ago I stood beside the old gravedigger on Memorial Day—he called it Decoration Day—and he looked across the monuments, the older ones leaning this way and that, obviously his favorites in that quiet yard, and he said, "This is *your* day." It was a gesture he believed in, just as he believed in keeping a record of everyone he buried.

It's been some time since the day aroused many sentiments, and a lot of people, thinking of the capable pleasures of visiting folks and eating outside on such a day, haven't much time for mortal thoughts. It's predictable behavior, and outside my window the soft maple has just come into its own and the spinach is up. Mortal thoughts are easily sidetracked in the face of such evidence to the contrary.

Life's a subtle courtroom, however, and the evidence never seems to be all in. It is true that across from the graveyard are just-planted fields and new lambs but while seeing that picture what was on my mind was the way the groundhog sounded in the woods when the old labrador chased him down, one high, ragged, unforgettable note, and the rattle of two dead walnut trees in the wind. Contradictory evidence abounds, and life at times seems to be a circumstantial race between the defense and the prosecution.

The race that life properly is, however, is the race to get something done, some work that counts, as a way of saying we've *been* here. "This hobble of being alive is rather serious, don't you think so?" says Thomas Hardy's character, Clare, in *Tess of the D'Urbervilles*. There's a moral law behind the gift of energy, although admittedly it appears to be a subtle one and is not found on any of the statute books.

My neighbor Mr. Taylor says that when he was a young man beginning to farm his ambition was to one day put gates in all the openings on his farm, and have those gates on hinges. Given the nature of those years when Mr. Taylor was beginning to farm, I understand his ambition as a moral one.

"It was not finances that kept me from this ambition," explained Mr. Taylor. "I got all the gates in and on hinges, then time passed and the posts rotted. And there I was." Farming people, better than most others, have a sure sense of the high likelihood of going through unfinished. Each day has its own mortalities.

So we come down here to the cemetery on this day, contrite and sober, for the reminder just as though we needed it. It isn't a bad ritual as rituals go and my attitude toward ritual itself seems to soften each year. That's a mortal reaction, I'd say, and it amuses me some although I don't think I'll yet aspire to membership in the Masons.

I believe in one fraternity and that one is underfoot. It makes me consider carefully the heedless optimism of spring, the studied humility of ministers, and the alluring promise of regular employment. It's a fraternity which has never been accused of segregation or favoritism (although it is guilty, it seems to me, of nepotism) and while membership requirements vary, it's quite a large club. Sooner or later, everybody gets in.

Confronted with membership, we'd all like to get in with dignity, perhaps with style. The gravedigger tells me about the passing of Phillip Lemar, someone he admired. "He was dying, there across the creek," recalls the gravedigger, "and Reverend Trout walked in. 'A word of prayer?' he asked, and Phillip Lemar said to the reverend, 'Well, damn it all, yes, if it'll do you any good. . . .' "

Those were his last words, the gravedigger said, and given some margin for misinterpretation, I'd say they had both dignity *and* style.

A Quiet Day with Geese-Watching

On my way back from buying the Sunday paper, I passed by the beekeeper's house and noticed him and his wife using their Sunday paper to mulch the grape arbor. This didn't seem an unreasonable use of the day's news, but I stopped to watch anyway.

The beekeeper's wife said it was their twelfth anniversary, which they were celebrating by mulching the arbor and working in the asparagus bed. All in all, it looked like a pretty big day of celebrating. Not feeling up to such festivities myself, I went with Miss Pinkie and Mr. Bobo to the pond to see what the goslings were doing. Earlier in the morning, the beekeeper had to rescue them from the cows who, filled with idleness and curiosity, spotted the goose family out for a stroll and stampeded them all over the pasture.

The goslings, four of them, were hatched out in a nest on a small peninsula in the lake, covered with tall grass protecting the family's flank. "Duck peninsula with multiflora tanglement," said The Squire, who had driven across the road in his truck to observe.

The Squire was a bit jealous because the geese had selected

the beekeeper's pond rather than his. The Squire had already been spotted in the clover field beside the pond, making duck noises, and the beekeeper accused him of using a set of "mallard postcards" to lure the geese over to his pond.

People in the neighborhood tend to keep up with each other in the spring. "I got *six* geese," one might say proudly to another. The Squire was only so jealous, however, because the geese were fickle and came and went, honking their way back and forth from his pond to the beekeeper's.

Last year, after hatching out at the beekeeper's pond, the geese led their offspring through the beekeeper's yard, and headed for The Squire's pond. The beekeeper's wife stopped what traffic there was on the road while The Squire made a space in the fence for the geese to get into his pasture. The beekeeper's wife said The Squire, at that moment, even looked a bit like an old gander.

The geese were staying put this year, though. Perhaps it was because they had good neighbors. About a foot from their nest was a pair of mallards who had a nest of their own. The beekeeper was sure that before the duck laid her eggs, she had been taking turns with the goose, sitting on the goose eggs. No one seemed to think much about this arrangement, and the goose appeared glad to be able to get out of the house for awhile.

When Miss Pinkie and Mr. Bobo and I got down to the lake, we decided to see if we could get a closer look at the duck nest. A few moments later, we were within a few feet of the female, who stayed on her eight eggs and let us know she wasn't to be trifled with. She fixed us with her bright, panicky stare, swelled up and made a noise like a small weightlifter. We backed away, assuring her we were just passing through.

I got back to the house and opened the paper but reading it seemed too much of a chore, so I folded it up, stuck it in the woodbox, and went out and sat on the porch. The goslings had escaped the cattle, the duck was hatching, and the bee-keeper and his wife were peacefully celebrating their twelfth wedding anniversary. I just didn't think I could manage any more news.

Gardening, I

My neighbor Mr. Lamke thinks he goes into his garden each spring because of his genes. He surmises that he has been genetically affirmed to grow lettuce. Mr. Lamke would agree with Charles Dudley Warner, who contends that the love of dirt is among the earliest of passions. Mr. Warner says that no matter how small a man's patch of dirt is on the surface, it is four thousand miles deep "and that is a very handsome prop-erty."

Theories abound as to why man loves a garden plot, and history is full of people enjoying a passionate relationship with dirt. The emperor Diocletian took leave of his empire to gar-den and was reluctant to return, finding great joy in presiding over his tiny kingdom of vegetables. Plato taught in a grove. Monks meditated in gardens. Jefferson was a gardener and Thoreau once grew seven miles of beans, asking, "What shall I learn of beans or beans of me?" (One of his conclusions was that a man could be too busy with his beans.)

One writer suggests that our passion for the growing of things is strong because we all came from a garden. "Man's first state of innocence was spent in a garden," he says, "and perhaps those who garden are trying to find Paradise again." I suppose there might be something to this theory although it is certain gardens are not innocent anymore because God has given us the cabbage moth.

Mr. Lamke says God has given him both the cabbage moth and the mole, a double affliction in the fall from grace. "I used to think," said Mr. Lamke, "that gardening was a battle between me and dirt. Then while I was fighting the dirt, I discovered the winner was the mole. Now it is a battle between me and the mole. The other day I put turpentine on my corn, to keep the moles away. Suzy said why don't we try that. I said, fine, where did you hear about that? She said O it just seemed like a good idea. . . ."

Like Thoreau, who had groundhogs in his bean patch and came to regard what they ate as a sort of tithe, Mr. Lamke and his moles coexist and his lettuce takes his mind off the moles. He has four kinds of lettuce in his garden and he reports that they all get along very well. He also says he has a head of lettuce he could climb on. "My garden," he said, "is a picture of serenity."

There are specialty magazines for gardeners who have such problems but none of us pay much attention to them. We had rather ask another gardener. One of the more delectable fruits of gardening comes in the conversational harvest. I do not know yet how turpentine on the corn is going to work, but I expect a report any day now.

Another man I know was having trouble keeping the birds out of his cherry tree, a problem he solved by buying a used burglar alarm from a bank and installing it in the tree. Now, whenever birds land among his cherries, the alarm goes off and his crop is saved. I suppose such remedies may be discussed in the magazines but it is more fun hearing them firsthand.

The magazines bother me a bit because they're such pious

enterprises. (Cato said that the profits of agriculture were particularly pious but he meant "pious" in the sense of "just." He
might have been thinking of the grower's struggle with the
mole when he said "pious" but it seems certain he was not
thinking of the middleman.) There's one publication, aimed
full-bore at the organic gardener, of which I am mostly one,
and it is written in the language of a Baptist missionary bringing the word of artful cultivation to the Dark Continent.

This is the magazine sermon I call "transcendental vegetarianism." It is not the worst sermon out in the world but it's
still a lecture and lectures give me a case of Trainman's Head-
ache. I listened to a lot of lectures when I was a child and all
they did was make me stay unhousebroken. Those people have
a second publication, about health, and a friend of mine
describes its tone as "one which suggests vitamins will cure
everything from hidden stop signs and split ends to a bad case
of the jilt."

There are no doubt large groups of pragmatic souls who go
into gardening each year in the way Oscar Wilde accused
Henry James of writing fiction—"as though it were a painful
duty." That is not my idea of it, however. I think sometimes
that nutrition is not even the idea of it and that I grow a garden
purely for the aesthetics of it. I like the way a garden looks,
and in certain frivolous moments, see it as one way of creating
order out of compost.

The best meal I have had this summer did not even involve
a garden. It involved a ménage â trois among pan-fried bluegill, cat-tails in fresh butter, and my appetite. The bluegill
came from the beekeeper's pond, and the cat-tails came from
a deep ditch along Squire Williams' set-aside. The appetite I
made myself, and the finish on it proved me a craftsman.

I do not know why people garden but the passion doesn't

seem particularly misplaced, as passion often does amid the curious range of a person's hobbies. A man near me told me he wasn't going to have a garden this year because it just went to waste. It is true, it did go to waste, right after his wife filled the freezer, canned sixty quarts of green beans, and the neighbors came in and took all they wanted. The problem is, the man has a passion for growing things. His garden swells until it's a plot against the horizon. I was by his house recently and noticed he has a garden after all, just as I suspected. I've seen men down with a passion of one kind or another before, and I know a backslider when I see one.

Gardening, II

In the road yesterday while hauling a load of manure from Mr. Ellis's stables to my garden, I passed a man hauling a load of manure to *his* garden. I didn't recognize this man, but we waved in passing, co-conspirators in the spring garden plot.

It is the time of year when we come to terms with vegetables again, a ritual of succession (in the case of asparagus) and maintenance (in the case of cut worms). I know the process as a benign, even beatific one, but I would not go so far as to say it was an epiphany. Perhaps it is merely that there is no backtalk from cauliflower, and we gain what it is that we gain by being high lord in a fiefdom of lowly vegetables.

John Cowper Powys, however, said, "We have no reason for denying to the world of plants a certain slow, dim, vague, large leisurely semiconsciousness." I pay obeisance to that sentence and in moments of unbridled optimism one might attribute a leisurely semiconsciousness to man, also. I do seem to recall conversations with broccoli which have superceded cocktail patter of the evening before.

Being faithless in the larger events, I am hard pressed by vegetables, too. The seed is so simple in design, so unadorned, how then does such an ornament as the brussels sprout come to us? How does the beet in its marrow know it is a beet? And does the lima wish to be a radish inside its prosaic lima heart?

I fret through the first few weeks as though labor pains were attendant to the sprouting potato, rather than back pains. I scrutinize the garden each morning as though I could foresee the future in its perfect rows. This is the syndrome The Squire refers to as The Dance of the Faithless. "It's a spring dance," says The Squire. "You see this fellow out in his cornfield, walking and bowing and sticking his hand in the dirt. Oh, in the spring, you can see this dance in most nearly every field. . . ."

The seed, with a compulsion of its own, goes about its mysterious business, shouldering the dirt aside, the fledgling green beans coming out bent-backed, wearing the garden soil like an overcoat. It's an admirable force, this quiet but peremptory compulsion. I've read that a curious-minded New England farmer attached a squash to a weightlifting device that had a dial similar to a grocer's scale to indicate the pressure of the growing fruit. Days passed, the farmer kept adding counterbalancing weight, and found his squash exerted a force of five thousand pounds per square inch. I subscribe to the notion of quiet compulsions, resting my case with the farmer's heavyweight squash.

As a compulsion, the seed itself is nearly matched by the gardener's own obsessions, conscripting all available space into his vision of a plot bursting at its seams with all manner of vegetables.

A while back, Mr. Taylor put in his garden, then discovered he had overdone it a bit. So he went down the road to Mr. Lundy's. "I've got too much garden for one old man," he said, and asked if Mr. Lundy wanted to share it.

Mr. Lundy thought that was a good idea, so he and Mr. Taylor worked on the garden together. I saw Mr. Lundy some time after the partnership had been struck and asked him how things were.

"Now," said Mr. Lundy, "we got too much garden for *two* old men. . . ."

Compulsions of this sort are problems, no doubt, but they tend to keep the gardener confined in his own backyard and his songs of compost and cabbage moth are thus kept to the level of background music for the tunes to which the rest of the world dances.

A Slight Sound at Evening (and Again, at Morning)

Over a hundred years ago, Thoreau wrote: "A slight sound at evening lifts me up by the ears, and makes life seem inexpressibly serene and grand. It may be in Uranus, or it may be in the shutter." That is a magnificent sentence, but if Thoreau had lived longer, it would never have occurred to him. Today, the shutter slams constantly in whatever contemporary wind there is, and the sound from Uranus is a refracted microwave transmission of a group of rock singers from Los Angeles.

I've read that the noise level around us is rising perhaps a decibel a year. This does not seem to unduly excite anyone but perhaps it is because no one is taking it at ear value. Speaking as one who has been accused of awakening at the sound of a plant blooming, it is dire news.

I seem to recall that Schiller, the German poet, used to rush angrily into the streets, provoked by the crack of a buggy whip, and that one of Tolstoy's characters stuffed paper in all the animal bells on his estate. These are people I see ear-to-ear with, and these eccentric notions regarding noise as an invasion of privacy are still leading ones in why I like living in the middle of a cornfield.

It is likely that most of the time the sounds here are not a lot different from the sounds Thoreau heard. The sounds of big commerce in the cornfields around me take place only two or three times a year and the rest of the time the fields are presided over by two rusty windmills whose low noise has become company. The only other disturbances that come to mind are the old labrador flushing quail in his sleep, and the beekeeper driving in the yard to yell something or other about being out of beer.

As adaptive people, we seem to take noise in stride (although noise is more often strident than striding) and the reactions to it vary. Mrs. Peterson, who moved from the country into an apartment in town, kept hearing snoring one night recently. She had not noticed it before, and thought the walls could not possibly be so thin. The next morning she was relieved to discover a man sleeping under the rose bush outside her bedroom window.

I also knew a man and his wife who lived across the street from the railroad tracks and every morning at 3:30 a freight sped through, causing all the furniture to shift around in the house.

This never seemed to bother them, then the fellow retired

and they moved to the country. The wife said the first night, at exactly 3:30 A.M., surrounded by decibels of silence, the husband bolted upright in bed and yelled, "My God! What was *that?*"

I have for some time envied those who can sleep through an avalanche of pianos, marking up this blessedness as the result of a clear conscience. Actually, I regard sound sleeping as dangerous, possibly the result of a great naïveté regarding the world's sinister intentions, and privately suspect that if anybody went to the trouble of making a survey we'd learn carthieves slept nine hours of dreamless sleep each night they weren't working.

Because I am sometimes given to grousing about what I consider an excess of noise ringing around in the rafters of the planet, I am often regarded in the manner one regards a person who grows a lot of small, fuzzy plants or keeps too many cats.

My case seems even more arrogant because I offer no apologies, regarding a certain amount of quiet as constitutionally given and, like the French proverb, I do not much like noise unless I am making it myself.

The Weather in Gurneyville

Country weather occurs upon both cornfield and hatbrim yet more upon the tongue. Here, low-pressure systems move through, curious air, old winters, and inclement seasons. Weather is the great leavener upon which country people rise to a sense of community. The talk in Homer Smith's General Store is always upon the weather. *Mare's tails and mackerel*

*skies, we are never at a loss for words, which pass among us like
proper favors.*

"In Jamestown," remembered the old blacksmith, "a
cyclone came and blowed a chicken into a jug. . . ." Worse
passed in the winter of '17: "Cows' breath," said the woodcutter, "condensed and hung in a great cloud over their heads. It
scared them so bad they stampeded through the side of the
barn. . . ." Heat in the summer of '21, 118 degrees in the
shade, the sky like a brass sheet.

Fronts blew erratically across Homer's old grocery counter,
himself an artifact in the archaeology of this store stuffed with
food, boots, nails, tools, ladders, buckets, ax handles, gloves,
seeds, and jars. Once the huge scarred drawers in the counter
held corn meal, salt, sugar, flour, hominy. People called them
the staples. Zimri Haines told his son Luther, that all a man
needed, store-bought, was the staples: "A little salt, a little
sugar, some kerosene." Zimri's grandson, Don, farming
hundreds of acres nearby with huge complicated equipment,
pondered this remark. "A man could do that today," he said,
"would be heard to smoke out. . . ." *And what about today,
Mr. Smith? What is selling? What's a fellow's line?*

Outside the store, the new corn ran in straight lines up to
the walls. The even green covered hundreds of acres. Gurneyville, store and five or six houses—is Rendel Carey in Gurneyville or in the country?—appeared like a surprise. People were
often drawn to stop.

> *What is this little place?*
> *Ever hear of Gurneyville?*
> *Nope.*
> *You have now.*

It is a picture they have of a life they think they may know about. Sit behind the counter in the swivel chair beside the rolltop desk. Put the feet up and believe a different life. Could this be what Homer sold? The pest man, drinking a Coke, knew better. "I've heard stories about those days," he said. "I hear it wasn't all they say it was. I hear horses were worked to death and men died too young." Homer's wife agreed. "I wouldn't go back," she said. "I have a dishwasher. . . ."

Once men began towns as casually as businesses. A town, in fact, was the collective image of business. A practical storehouse upon whose metaphorical shelving rested a variety of businesses. A hundred years ago Homer's grandfather began the store in one room of his house. Another room was the post office. Restless, he built across the road, an investment in hope: Gurneyville, named after some forgotten old Quaker, would be a town. There was blacksmith, wagon-maker, broom maker, schoolhouse. Homer's father bought out his father-in-law in 1893. Paid him thirty-four dollars for the stock. By that time everybody knew there would be no town.

There is an old photograph: C. H. Smith, groceries and hardware. The elder Mr. Smith is a bit stocky, a Chaplinesque mustache. He stands among the well pumps, beside a tired old man wearing a vest. Another man stands to the right, in front of a window sign advertising Foley's Honey and Tar, "a great throat and lung remedy." The second man is formidable in appearance, full black beard and wide hat, dark and brooding. *C. H. Smith. Dealer in groceries, boots, and shoes. Merchant's Gargling Oil Liniment and Worm Tablets. For man or beast.*

Salt in 280-pound barrels up to McKay's Station on the Grasshopper Line (so slow, they said, it never frightened the grasshoppers off the roadbed), barrels of Baldwin apples, and bananas a treat from Cincinnati. Crackers in tin boxes, salt fish in round barrels and dill pickles in the brine with a suitably long fork.

The storekeepers, three generations now, stoically regard the convolutions of the brash new century from their crossroads. Automobiles appear on the mud roads. ("They ran along but nothing was pulling. It seemed . . . wonderful. Beyond us. No one understood very much. . . .") Soon, Homer is running the huckster route in an old panel truck, fifty different miles each day, cracked corn and bran millings in one-hundred pound sacks on the front fenders, coops of live chickens in the back, gingham in bolts, kerosene, and grocery orders. And eggs. Dozens of eggs. As many as four hundred dozen. Traded by every farm wife on the grocery bill. Everyone has chickens. Did not Bernie McKay once drive to the Grange meeting with a row of old hens still roosting on his fender? Homer tilts through the countryside, careful with the eggs on aberrant dirt roads. He stops by and gives Vossie Hackney a haircut, and back to Gurneyville in the failing afternoon.

The store persisted although the town did not. The school closed, too. The schoolhouse was in the northwest corner of the crossroads. Homer bought it and moved in. He never liked school. He knew he didn't like it before he ever went. He said he wouldn't go. His parents were of another mind, of course. The older kids picked on Homer. He felt, somehow, *inferior*. The lessons were monotonous. As soon as he could he quit and went to work in the store. Then he bought the old schoolhouse.

In the twenties, the elder Mr. Smith said to his son, "I have treaded these boards as long as I care to. You want this place?" Homer, who had no other place to go, took over. His

father walked across the street where he could sit on the porch
and watch who went into the store.

Things narrowed down during the Depression. People put
their cars up on blocks, unable to buy gasoline. "The banks
went under in Wilmington and we'd get checks on them. 'Why
didn't you cash that check sooner?' they'd ask me. Out of sorts,
they were. Things were awfully tight. Everything was so cheap
but there was still no money. The township could issue a grocery order for someone really down and out. Just the necessi-
ties, however. Corn meal, soap, flour, a little jowl bacon. The
cheapest cut. We sold a lot of jowl bacon during the Depression. . . ."

Business persisted, however. Like men and agriculture.
Homer replaced the old icebox with an electrically refrigerated
meatcase. He was skeptical of such a thing at first. He understood ice but not electricity. It seemed to work well enough.
The huckster route continued, too. Farm wives were not yet
released by the automobile, appliances, and convenience foods.

Homer had seen all manner of changes but he had little to
say. The changes seemed to be mostly . . . *outside*. There was
not much to observe in Gurneyville. Its proportions, narrowly
circumscribed by endless cornfields, had not changed in a
hundred years. In spite of the shopping centers, people still
found their way to Homer's store. With some of the older customers it was perhaps nothing more than keeping faith with a
system which was comfortable to them. Clive Lundy, nearly
ninety, still drove over from Ogden to buy his chewing tobacco.
His younger brother, Frank, who lived down the road, stopped
by for milk.

One day the health inspector came by and told Homer
about the new regulations. The new regulations said the store

had to have running water and a bathroom or Homer could not sell fresh meat. "Not at my age," he said to the inspector. He did away with the meat counter and covered up the old butcher's block.

At noon the farmers came in and bought soft drinks. They sat in the swivel chair and leaned back against the old safe that Homer never locked. Even Homer's grandfather never kept any money in it. People never believed this, though. Someone was always trying to get into it. Once burglars blew off the door with nitroglycerine. They made quite a noise in Gurneyville. People are surprised when they hear the store has been broken into so many times. "How does anyone even know where Gurneyville is?" they asked.

Homer's father fired a shotgun at some burglars one night and they left a trail of tools across his grandfather's hog lot. Once thieves crawled under the store and came up through the floor. When Homer opened up he thought it was ground-hogs. Until he noticed the stock missing. After one break-in, there was a note on the door: "It was my husband who broke into your store. He beat up on me so I'm telling."

Farmhouses as well as the store were burglarized. Country people were no longer immune to fear. Theirs was perhaps even a stronger fear. New people come to the country seeking a way to be. Come to the country in a tin box with wheels, plant yourself on five acres of a grainfield, expect a new season. Country people, too, will have their evolution. And fear in that, even with the turning. This was the change Homer noticed: The farms now were huge, and in between were the subdivisions of people come to live in the country that was becoming neither country nor town.

On Homer's seventy-fourth birthday, he celebrated by

closing the store at six o'clock instead of nine. The lack of exercise after supper, he noticed, had caused him to gain two pounds. "Lots of hours in a day when you get up in your seventies," he said. "Now I take one day at a time. I have no definite plans. I'd rather be here than elsewhere. I've never been anywhere else. Sitting around the house, how would that be? When Mudge Blair sold out in New Burlington he nearly went nuts. Went up every day and looked around. Finally, he bought the place back. . . ."

The milkman stacked the shiny cartons in the cooler. A woodcutter's wedge held the door open. "Twenty years ago," said the milkman, "there was four dairies in Wilmington, twenty-two little grocery stores. Eighty per cent of the milk was sold door-to-door. Now it's ten per cent. Someone was at home once, you see. Knock on any door today and no one's there. Four independent stores left. I don't know of any general stores. I don't want to say this in front of Homer but when he goes, no one will come in here. There's new regulations and no future."

"Oh, maybe someone will buy it," suggested Homer. "There's always someone ready to buy anything." I wouldn't recommend it to a young fellow though. Not if he had a good job. . . ."

The milkman leaned back in Homer's chair. "Homer," he said, "we have seen many things in our time."

"From the buggy to the moon in my period," said Homer. "That was my time. Lot of new things about. What to do with them? I used to take a ride on Sunday, go back over the old routes. I knew everybody, stopped at almost every house. Most are . . . dead. Lots of new places. Makes me feel I'm . . . old. I don't ordinarily think that."

The floor yawed this way and that under the men, a sea of old planking. The screen door slammed. Bees droned outside the window.

"Homer," said the milkman, "why keep on at your age?"

"Why, Carl, the exercise. I need a little exercise. . . ."

Homer went on until one day, like his father, he decided he'd had enough, and he locked the door and went across the road to sit on the porch, only he had no one to watch.

He sold the stock and kept nothing but the gas pump, which he used for himself. The building echoed. In the bad months he went over and, for the exercise, walked up and down inside. Around him, in the late spring just after his seventy-ninth birthday the new corn, in perfect rows, pushed up to the back of the old store as it had for most of the century. "You have to quit sometime," he said. "That time finally got here. . . ."

Let Us Now Praise Famous Sharecroppers

I have always looked upon Adam and Eve as sharecroppers, victims of the first tenant-landlord misunderstanding, and what came after was *work*, as a curse. What we had there was a parable for the evolution of garden to field crop. There is considerable difference. If I had been working on a book as ambitious as that one, I would have had the confusion of tongues occur not at the Tower of Babel but in a cornfield. As a countryman, it seems to make more sense.

Ever since the auspicious debut of agriculture as punitive measure, work has been both anodyne and affliction. We've struggled with work and, of late, with the definition of work which might be the hardest work of all. "What *is* work?" asked

Mrs. Shidaker as we sat on her front porch one evening. "Is it something devastating, with no pleasure to it? On the farm I made fifty pounds of butter a week. That was *work*. I got a little weary of butter. . . ."

Mrs. Shidaker thinks that once the farmer did the hardest work. "My Eddie was up at four, to bed at nine. In between, he ordered the fields. We had two mules and a plow that turned a single furrow. The mules did not have a cab with air-conditioning although they could sound in stereo at dinnertime. Eddie plowed, harrowed, and disced, attending to the soil. If dinner was not quite ready, he had a few minutes to read the mail. . . ."

I've a neighbor to the east who recalls hoeing out by hand one hundred sixty acres of corn. The father of a neighbor to the west put drain tile in ninety-seven acres with a ditching shovel. East, west, the balances of that time. Helping another neighbor work on his cistern, I dug a ditch eight feet long, four feet deep. Up to my waist in unmanageable clay, feeling my own tongue begin to confuse, I imagined the work done in the short history of this country and found it unimaginable. The Ohio Canal was a ditch three hundred and nine miles long, forty feet wide, four feet deep, dug by men with shovels who worked for thirty cents a day. "If one hasn't a horse," wrote Van Gogh to his brother, "one is one's own horse."

Grass grows in the Ohio Canal now and if there is a particular wisdom ascribed to the hoe over the chisel-plow, I am slow to recognize it. What's the wisdom of sacrifice if it doesn't hold? If the slave died under the stone of the pyramid, then the entire thing became his monument. What to make of the fact that the loveliest things made have often killed the makers? There is, indeed, a knot in the drawstring of the universe.

In a tool shop, working summers to buy college, Shorty

worked beside me, singing, "Today's Thursday, tomorrow's Friday, and the next day is SATURDAY NIGHT"! No love there and he, unlike my privileged self, was in for the duration. After a tour of the tool shop, I chose my father's hayfields. Unloading eighty-pound bales of alfalfa under a tin roof where the temperatures approached one hundred twenty degrees, I looked at the fields with a near-equal shop dislike, but I was not afraid of them. I *understood* the fields.

Later, as we think it befitting an education, I came inside, washed up, folded myself beside a desk. One of my memorable editors wore a brown shirt when he fired someone. The staff called it his Gestapo shirt. His madness was institutionalized by *his* editor, a pious man who washed his hands under Picasso prints hung carefully on his walls. As I recall, I was fired under a Picasso print. That was no place to work. That was no place to *visit*. And it was Picasso who said once, "Work is a necessity for man. Man invented the alarm clock."

But who *said* work fifty weeks, take two off? And was John Henry a steel-driving man or a steel-*driven* one? There's a difference worth mentioning. That's the way the mind works after an instructive moment under the Picasso prints, after throwing away the clock and going after time in one's own way.

The clothier in our town buried his wife and was first to leave the funeral, hurrying back to open up the store. "I smell groceries everywhere I go," says ex-grocer Mr. Scott. "You got to have the glad hand and the sweet hello." And when I asked Mr. Mendenhall, an old farmer, would he do it all over again, he aid, "I reckon I'd have to. . . ."

Thinking about work, I've come down to this: people work because it is, by and large, all that they *can* do. Work is less boring than the alternatives. What else can a person do for a sizable portion of his waking time? Can he eat or otherwise pleasure himself for eight hours? It is doubtful he can even be amicable with others of his species for an eight-hour stretch although that, too, has become a modern craft, under the name: *counselor*, or perhaps *mediator*.

Such professions are mostly unknown in my neighborhood because it is a neighborhood of simple occupations. There *is* a lawyer down the road but people seem to have forgiven him this trespass because he has eighty acres of good beans and his fencerows are neat.

As for me, twenty years from the hayfields, I am back working with my hands. That is, the hand as connected to the eye. The judgment won't be in for some time. It is one definition of craft and that is not to say I do not, upon occasion, dream pleasurably about the cartharsis of working with baled alfalfa under a tin roof in August.

What's the hardest work? To work for one's own self.

Auctioneer's Song

The man was an entertainer, that was why. He was a man doing a high act off a hay wagon with nothing but his own voice to catch him. As for the voice, it was a voice that could make a ballad out of moldboards and buckets of bolts and the fellow out in the yard would feel like he'd just bought fifty acres of bottomland cheap. It was a voice that could talk a woman into a weekend or sell a man a swamp for a cornfield.

And what was more folk song than an auctioneer's voice? He was the man who wrapped up a family's belongings, cream pitchers to grain drill, and it was blues or band music, depending on the circumstances. He was the man who knew what everything was *worth*.

That was why Mr. Workman became an auctioneer. He liked the fellow's *line*. On a good day, he could have people thinking buttermilk came out of a honeydipper. On a bad day, well, on a bad day, he couldn't have sold a mild spring to an orchard man. And *that* was what he liked: good days as opposed to bad days and just enough bad days to give the good days flavor.

There were precedents in the family, too. His grandfather was a trading man, and his great-uncle. So was his father. "One thing my father could do without any embarassment," remembered Mr. Workman, "was cheat a man out of a horse. . . ."

The uncle, as a young man, went to the bottomlands with a horse. He threw up a terrace across the field and back, returned to the barn and unhitched.

"What's the problem?" asked his wife.

"God created this earth but he has not yet created a good enough breaking plow to tear it up," he said.

The uncle quit farming, started trading, and became wealthy. If such a thing does not register genetically, then it may be instructive in other ways.

So the then-young Mr. Workman went off to Reisch's School of Auctioneering, Mason City, Iowa. "Just for the refinement," he said. After all, he didn't want to merely stand up and make a noise. He learned how to breathe right, make his brain work along with his voice. He came back, said, "Boys, I'm a whale of an auctioneer. Put me up. . . ." And they did.

He sold pies at socials, worked free sales, anything. He was milking cows one evening and Jess Stanley walked in.

"I'm going to have one more partner in my auctioneering life," said Mr. Stanley. "You want to be him?"

"Yes," said Mr. Workman.

"I'll see you next week," said Mr. Stanley.

Mr. Stanley had two prices for a sale. Regular price and free. He was good and fast and honest. He started like Mr. Workman did: auctioned off an old tomcat in the front yard. Of course, auctioneers tended to be cut from the same bolt. A fellow Mr. Workman knew began selling when he was eight. Sold his own brother a duck, which belonged to a neighbor. Once, when a three-bottom plow fell off a truck in front of his house, he traded it away before the owner came back for it. Then there was old Charlie Bond. Charlie Bond would beat a fellow down in a horse trade, then send him a check the next day to make up the difference. It wasn't the money Charlie Bond liked, it was the *trading*.

Mr. Stanley taught Mr. Workman that a good man had a good sale. There was Frank Lundy's sale, for example. Good, clean items. Everything marked. Sheep pinned just so. Mr. Workman was proud of that sale. Not a box of bolts brought less than it was worth. Everybody left satisfied, buyer and seller. Good man, good merchandise.

Other things Mr. Workman learned on his own. He learned to develop his peripheral vision so he could see all the crowd. "If you can look straight ahead and read the newspaper at a right angle, you're there," he said. And he learned how to handle a crowd. He remembered an evening sale, nice evening, dark coming on. A young fellow kept bidding a quarter. "Twenty-five dollar item, he'd say 'Quarter!' Irked me no end. I said, 'When I was a boy on the farm, worked an old mule in the truck patch. He was hard to handle so I got him cornered and worked him over with a shillelagh. Dad caught me, said,

'Son, someday that jackass will come back to haunt you.' I never believed him—until today.' "

He learned there were some things he could not auction. "Fellow I knew bought a truckload of brassieres," he said. "Ten cents a piece. Couldn't go wrong. Good brands, all sizes. Hired me to sell 'em. Boy, ladies won't buy brassieres at public auction.

"It was my experience that a bigger cup was easier to sell than a smaller cup, but either way you were pretty much talking to yourself. Everybody just looked the other way. I bought them at a nickle a pair and put them in a little storeroom down here and I let the ladies come in as they wanted and finally sold them."

And he learned that the only way to fool people was to find a bidder who didn't know the merchandise. He sold a manure spreader to a woman once. She bought it at a thousand dollars. It was worth half that. Mr. Workman felt a little guilty afterward. "I wouldn't have given that kind of money. A city fellow was bidding against her. A fellow I let stand but a woman, well. . . ."

He told folks that when he was buying he was stingy, and when he was selling he always gave them the auctioneer's guarantee: "If you buy it and it falls apart, you get to keep all the pieces."

He loved to buy and sell. He said, "I'm like the fellow who, remembered as a bit backward, shows up at his class reunion only he's driving a big fine car. 'Well,' he explains, 'I buy at five dollars and sell at fifty-five dollars. That five per cent really adds up.' "

The sound of his own voice at work over a crowd pleased him greatly. It was the *rhythm*, not the words. He admitted that a lot of auctioneers didn't know what they were saying.

He thought of it as an *expression*. Sometimes he sold in a reverie, emptying the barn while thinking about fishing. Once, he sold for seven hours straight. Before big sales, he went out somewhere and worked for nothing. "Just to get tuned up," he said.

He saw auctioneering as an occupation of public trust. People trusted you or you didn't work long. "The auctioneer," he said, "was a respected man in the community. He was a knowledgeable man. He knew the value of everything and his respect came from that.

"I am an old-fashioned auctioneering man. The expression builds in me and I want to blow it out. I think of it as an explosion. I amuse the seventy-five per cent who come to be amused, and I sell to the twenty-five per cent who come to buy. . . ."

The Sound of Ingenuity

I seem to hear these days the sound of industry coming across the cornfields. I use industry not in the modern, technological sense but in its old-fashioned context, *ingenuity*, which my dictionary tells me is an obsolete use. I go right on in the face of this obsolescense, for such going on is a practice of mine and I hear what I hear. This sound reaches me at odd times, and slightly, the sound of *tools*, a kind of workshop-and-garage cadenza. Or perhaps, given as I am to following the odd music, I *think* I hear them.

At any rate, I normally enjoy pessimism, confident in the knowledge that misadventure is imminent, and my hearing usually errs on calamity's side. Thus I am pleased when I hear an old music in the air instead of the normal callithump of lost office, personal inflation, and the various unmet obligations.

This sound from across the way is the sound of *tinkering*, a pliant old craft fallen to disfavor in our specialized times. I lack the appropriate weighty statistics as to what constitutes a revival and so I document by listening and watching, two other pliant crafts on the edge of disfavor.

I date the small technology-and-tinkering movement in my neighborhood from the time Mr. McIntire chased the fuel oil man out of his yard. In the space of a few days he built a chimney and repaired a junkyard stove. Said Mrs. McIntire afterward, "We thought that at Christmastime we'd turn on the furnace for awhile and let the children see what it was like in the old days. . . ." After that, wood stoves, both traditional and innovative, ran through the neighborhood like a new minister.

Mr. Ellis made a box of copper coils for his fireplace chimney, circulated water through the coils and ran that through his air conditioner, which heated his house. Mr. Ames, by summer, had made his own air conditioner by using a car radiator and running well water through it. Mr. McIntire got disgusted with his garden tractor and said he was going to build one himself. "You can't build a tractor," Mrs. McIntire said. "*Somebody* builds 'em," Mr. McIntire answered, undeterred.

On the other side of town, a fellow I know had a dream whereby the Lord appeared in his tool shed and said, "Why can we not use a hydraulic motor. . . . ?" The Lord then diagrammed a round cylinder with a piston in the center and when the fellow asked what kept it from leaking, the Lord said, "The disc shall ride in the cylinder and the groove below will seal it. There shalt be no leaks." The fellow went to the kitchen table, drew some sketches of a car run by a round cylinder, powered by a hydraulic motor and a bank of batteries which were supposed to store enough excess energy to move the car without a fuel source. "Who am I to argue with the Lord?" this fellow asked.

My neighbor Mr. Ellis, an electrician and master tinker, says the man has invented a perpetual motion machine, the secret of which has remained elusive for quite some time now, but the fellow has himself some detailed blueprints, plans to have a machine shop build his engine, and Mr. Ellis and I sincerely wish him luck. Mr. Ellis doesn't like the idea of perpetual motion's arrogant refusal to yield to some dreamer's vision any more than I do.

These are some of the pictures, the view from out here, and I'm pleased if they contrast with the news networks' nightly tableau of humanity staggering under the weight of the eleven o'clock crisis. When I want the news, I wander out through the neighborhood, following the sights and sounds. The television and the papers bring us news largely in fits and starts, certainly not the whole picture, and like Mr. McIntire and the fuel oil man, I am no longer buying; I'm for making my own, hammering out headlines of my own devising on a bench in the garage.

It is possible that given the incentives of government and corporate affliction, the entire country may end up in the garage, inventing its own future. It's a picture I like a great deal, and there's some evidence for it. This was a country, after all, that tinkered itself into existence. Franklin was a tinkerer, and Jefferson and Ford and Edison.

My neighbor Mr. Hackney contends the army liked farm boys during World War II because they were ardent tinkerers, too. "Farm boys took a little wire and a bent nail," he says, "and patched things up. Later on, they became pretty good welders and machinists. They could take hold of something they didn't know much about and make it go. This was their nature. . . ."

Tinkering is a venerable form, and I think it is about to come into its own again. It is possible this is a time much like the early part of this century in Dayton where a group of local fellows, puttering around in barn and bicycle shop, invented modern transportation, perhaps a mixed blessing but innocently conceived in a burst of remarkable energy.

A fellow named Deeds, a friend of the Wright Brothers, built an automobile in his barn but decided there wasn't much future in it since over a dozen cars were being manufactured in Dayton at the time. So he and Boss Kettering, another engineer, applied their considerable skills to completing the auto, coming up with the electric ignition system and electric starter.

At his farm in the edge of Dayton, Deeds's men, in the cowshed, perfected the first V-8 auto engine. Arthur Morgan, who built the system of dams to protect the city after the 1913 flood, did hydraulic experiments in the swimming pool and Charles Steinmetz wired Deeds's garden, testing a German theory that electric current stimulated the growth of vegetables. (It worked, but the utility bill was enormous.)

A neighbor of mine, Frank Irelan, grew up in Dayton about that time. His father managed the company Deeds and Patterson started, and the young Mr. Irelan recalls sniffing the local atmosphere and being influenced by it. "We heard all this talk," remembers Mr. Irelan, "and we soon learned the nonmenclature of working things. This tinkering in the air made us inquisitive. Everything these people saw could be improved, or fixed, or altered. Their thinking was always, 'What does it do?' not 'What does it cost?' They were out to express themselves. I went out and sat on the dam near the old McCook Field and watched the Wright Brothers, came back home, and decided to make a glider. I got some two-by-fours, stole a bunch of mom's bloomers, and nailed it all together. I was trying to invent the lighter-than-air two-by-four, I suppose. I must have been about ten. Anyway, me and a friend of mine winched this contraption up on the garage roof and flipped a coin to see who would sit on it. I won. Or lost, depending upon how you look at it. He gave me a push and I went off the eaves and landed in the backyard, two-by-fours and women's bloomers all over me. God-dang. We were always going to the moon, you see. . . ."

And now we have been to the moon. Stuck at least temporarily within the confines of our own shrinking planet, we begin to take an accurate measure of the times. Around me, a rising diligence tempers the energy crisis. The crisis of *personal* energy is the only crisis to be feared. Meanwhile, I listen for the subtle allegretto of a new industry coming across the fields. Even I, sunny pessimist at best, may try a rusty, independent dance beside my woodpile.

Of Crowbar and Profanity

I have just finished tearing down an old barn, using the time-honored methods of crowbar and profanity. The barn, which belonged to a neighbor, had been damaged by a high wind and I brought the old building down on its sills as much by an artful use of curt Anglo-Saxon as by the crowbar.

I began on the roof, peeling off the wooden shingles. It looked to be a simple job until I discovered there were about a half-dozen layers where the owner had kept re-roofing the barn over the preceding layer. Future archaeologists, uncovering a section of this roof would, no doubt, be able to date the leaky cycles of consternation in the owner's life by each stratum of shingles. These layers, of course, made the roof exceedingly heavy. It suggested to me why the old barns contained much stalwart underpinning: to hold up the shingles.

I found that by standing underneath, on the top beams, I could punch the interwoven sections off with a long pole. It made a fine shower of shingles and I made a mental note of the wreckage as splendid kindling for the winter.

Soon the old girl was immodestly down to her underpinnings. (I say "girl" here because, as with the French, all things have a gender to me and old barns, like ocean liners, are feminine.) Modesty aside, she was quite a sight. The whole northern end was still in good shape, handcut ten-by-tens in the bents held snugly in place by flawless mortise-and-tenon joints.

I wondered if this was one of Frank Stanley's barns. Mr. Stanley, who built many of the old barns of my neighborhood, never build a barn with anything but hardwood pegs in the joints. He scorned nails because he thought the wind twisted them. Because of men like Mr. Stanley, the barn architecture at one time in the neighborhood rivaled the craftsmanship of the houses. In some places, the barn took aesthetic precedence over the house. There are still a few of these farms left near me. The magnificent barn stands quietly behind the farmhouse, in the manner of a family portrait where the pretty child is overshadowed by the lovely mother. I once heard of a man who spared no expense keeping his barn in immaculate shape but refused to put indoor plumbing in the house. He said it was too much of an investment, and besides, he spent most of his time in the barn anyway.

Sitting fifteen feet off the ground, thinking of Mr. Stanley's methods, I found myself given to a disconcerting theory. The theory was that the fine old architecture was not necessarily built out of taste and sensibility, but out of a lack of tools. That is to say, man was a proper craftsman when the hand was still connected intimately to the eye. As soon as power tools and aluminum siding came round, the craftsman used them, along with any other short cut, and soon found himself a union carpenter. There isn't anything inherently wrong with short cuts but they have something to do with explaining why most buildings today look like they were either poured or assembled, and the human hand vacant from the premises.

It was a theory too sober for such a day so I laid it aside, climbed down, and cleared off the debris on the barn floor. After that, I removed the planking, exposing eight, thirty-foot long, foot-square, handcut floor joists. Sitting on the balcony of the top beam, I surveyed the scene. I was through the prologue and down to drama.

At this point, I called for reinforcements. I drove into town, stopped by the college gymnasium and hired a stout fellow as my last-act assistant. He told me he was called The Hog, a sobriquet he had earned for his skill at imitating boars in various states of distress. The Hog, I learned later, was a linebacker on the college football team and his teammates contended that on the playing field, The Hog's noises in a pile-up were most discomposing. He was a pleasant fellow, from the farming country up north, and he looked as if he could shuffle pianos.

We drove some of the pegs out, used a chain saw on the other joints, and soon the old framework had slid gently into a pile of large sticks. Mr. McKay came over with his backhoe and, calmly smoking his pipe, swung the big, untenable beams into the back of his grain truck.

I like working with Mr. McKay because he is imperturba-

ble when a project's eccentricity arises, as it usually does
although we went through the afternoon relatively free of it.
Mr. McKay farms with his brothers and their father, remain-
ing unruffled even though he is the only one not partial to
Ford trucks, a source of much supper-table litigation. Mr.
McKay prefers the Chevrolet truck.

Not long ago, he mired his Chevrolet in the edge of a
nearby field and a couple of neighbors came out to help. One
backed his pickup into the edge of the field, up to the back of
Mr. McKay's bigger truck, hooked on with chains, and they
both revved up. Nothing happened. Mr. Steinbarger, one of
the neighbors, looked down and noticed that while the pickup
was going forward, so was Mr. McKay's truck.

He walked over to the driver's side where Mr. McKay sat,
one arm in the window, while the motor raced and the truck
sank deeper into the mud.

"You ain't got your truck in reverse," said Mr. Steinbarger
."Nope," said Mr. McKay.

"Why not?" asked the puzzled Mr. Steinbarger.

"Ain't no *Ford* gonna pull me out," said Mr. McKay, taking a puff on his pipe and indicating with a nod the pickup
spinning away unprofitably in the other direction.

That is Mr. McKay, a man who keeps his own counsel.

Soon, The Hog and Mr. McKay had the beams in a neat
stack in my own barnyard. Such beams they were, too. They
had once held up a farm's traffic and commerce, and were
prepared to keep on doing it. They were timbers to stand under
during a heavy afternoon. They could support a man's varied
prospects and set a fine example for holding to in high winds
and other rural disorders.

We sat on them for a few minutes and everybody seemed
pleased. The Hog seemed pleased he had been able to lift one

of the beams by himself, Mr. McKay that he had maneuvered the timber so adroitly, and I was pleased that I'd undertaken the whole thing.

Outside of using a piece of handcut beam for a fireplace mantle every now and then, there isn't much use around here for the big timbers. I'm not sure I have a use, either. I think of putting the beams back up as a house, a sort of Phoenix-house rising from the disgruntled ashes of demolition, but I can't say if it will happen or not. I put up three essays last week and not one of them had Mr. Stanley's mortise-and-tenon joints. All I know is that when I see one of the lovely old barns sitting idle, I have an irretrievable urge to reclaim it, even if this urge comes to nothing more than stacking it in the back-yard.

Just yesterday, I saw an old barn abandoned in a field, the rows of corn running right up to its weatherboarding, and before I could stop myself I thought: there's a barn I'd like to tear down.

At such a moment, I seem to reflect the schizophrenia of the day, trying to balance the attitudes of destruction and redemption in the same act.

Chester Township Boys Go to War

It is not known precisely what propels men to war. These things are suggested: idealism, conformity, fear, boredom. In curious and collective proportions these things are known as *patriotism*, or *religion*. Thomas Haydock the tanner's son followed his brother John to war. He did this by running away in the night, hiding in Jesse Hill's cornfield, and lying about his age.

He was fifteen. Miss Anita Weeks, of the neighborhood, said: "The pretty boys have all gone away. . . ."

One half the adult male population of Ohio went to war, one hundred ten of them from Chester Township. Although the villagers boasted that no one had ever died a violent death in New Burlington they were knowledgeably acquainted with violence. In the village chronicles were stories of high risk performed in uncommon times. The Haydocks were related to the Howes who fought on both sides of the Revolution. The old Quaker farmers, the Comptons and the Waltons, abandoned the fields for the War of 1812. The Manns, who settled near the covered bridge, were German mercenaries in Queen Anne's War. The McKay's, whose name itself meant "son of the impetuous one," could trace their history back seven hundred years, most of them filled with Scottish bloodletting.

In a history as recent as the early 1800s these families fought the forest, wrestling it into cornfields. For nearly a hundred years they fought the remnants of Indians, reluctant fields, heretical seasons. But men are not decorated for the order of their fields. When Enos Finch, William Collett's paid substitute, was killed in the Civil War his neighborhood held him responsible. Those who remained behind to farm were suspect, although there was no logic to such casuistry.

Warren Shidaker was seventeen when shot in the head during the Battle of Peachtree Creek as Sherman marched to the sea. This was the three-hundred mile long, sixty-mile wide march with sixty thousand men which led Sherman to say: "I am sick and tired of war; it's glory is all moonshine." Sarah Shidaker recounted family history: "Warren went away to war in early 1863. There was a candy-pulling in his honor. His little sister Lida could not stay awake though she tried and the next morning by her bed, beside the brass candlesticks on the two-drawered cherry table, there was a plate of taffy Warren left for her. He was killed July 22, 1864. Solon Carroll saw him killed and buried him in a blanket. After the war Mitchell offered Solon and Alfred Van Tress a hundred dollars each to take him after the body of his son. They brought it back and buried Warren with his family, Clarrie and Kate and Riley and an infant son born dead. Mitchell planted a ginkgo tree over the grave. 'I wonder if it is our son,' Mitchell said to Betty. 'It makes no difference now, Mitchell,' she said. 'It is *somebody's* son. . . .'

"On a spring housecleaning many years afterward, Lida and her daughter, Elizabeth, found Warren's army overcoat and cap which had hung in the attic for a generation, riddled with moths. There was nothing left but to burn it although when the blue wool went up in smoke and flames out in the backyard by the ash hopper it was to Lida as though they were cremating Warren himself. . . ."

The township boys marched across a strange land. Elias Dakin Harlan said: "I was on my own two feet and saw horrible service." He fought under Garfield with the fortieth Ohio Volunteer Infantry, at War Trace, Chickamauga, and Moccasin Point. Cut off and surrounded in a barnyard he and two others surrendered. The Confederates opened fire anyway. One was killed and Silas Hawke, a Harveysburg carpenter, was taken prisoner. Elias feigned death. "No one lay as close to the ground as me," he said later. "I became lifeless, and felt myself flow into the sod. . . ."

Silas Hawke was a prisoner for fourteen months, eight of them at Andersonville where men lived on a pint of corn meal a day and died of exposure, starvation, and disease, taken away in wagons to a common grave. Silas escaped as Sherman pressed to the sea and came home at suppertime to smell ham frying. Not recognizing her husband she thought he was asking for a handout. His young son, Victor, whom he had never seen, played on the kitchen floor. The stranger, Silas Hawke, began to cry. Later, he built substantial houses and was known as "a man of intelligence, good meaning, fond of reading and conversing."

William Miller of the same regiment was also taken prisoner. He was sent to a Confederate hospital where a nurse placed a gangrenous poultice on his wound, causing his death which was neither quick nor merciful.

After the war the village soldiers were wounded, dead, healthy, even married. John Blair, the first township man to enlist, at the first Battle of Bull Run and soldier with the Army of the Potomac, took a ninety-day wife in the south, a strange but permissable custom of the time. Finding the custom not as permissable, the new Mrs. Blair arrived in New Burlington before he did. They lived together. Later they adopted a small child who became a musician. A villager remembers the day the little girl met the Blairs: "The child was Ida Clipperd whose father choked to death on a piece of beefsteak. She came into the village. It was raining and her blonde curly hair was tightened to her small head by the water. As she crossed the creek there was John and Lizzie Blair. They took her on their knees. First one then the other. She never left."

Thomas Haydock returned to become the village cobbler. "No woman ever looked so splendid as my mother," he said upon his arrival back in New Burlington. His brother John, a blacksmith who shod army mules with the Second Ohio Heavy Artillery, returned injured severely by a mule's kick, and deaf from the roar of the cannons. His wife wrote him notes. Because they had been army staples he never again touched rice or tea.

Three black ladies came to live in the village. A villager recalled stories: "They were sisters, lovely women of the old school. One had a child. His father was their master in the south but no one ever said anything ill of them. The child grew up in the village and became a cobbler. . . ."

Afterward, an old Confederate soldier moved into the village. His name was David Ricketts. "Thomas," he said to the cobbler, "there is a line of demarcation between the white man and the damned nigra." He made coffins for the undertakers who considered themselves tolerant men.

For years afterward the brass band practiced for one major occasion: Memorial Day. Thomas Haydock played the tuba. His brother John played the snare drum. His son Trevor, the baritone. Fifty years later his daughters cannot hear the baritone without thinking of him. On Memorial Day they marched in white to the cemetery where Sarah's feet swelled in her new white shoes covered with dust. The villagers carried bouquets gathered in Frank Hansell's runabout the day before and kept fresh in tubs of water at the church. The band countermarched in the cross streets and broke step on the bridge for fear it might cave in.

In the next century Sarah had a son of her own named Warren who lived in Texas. Visiting in her daughter-in-law's dining room under a steel engraving that belonged to the daughter's grandmother, Sarah considered the engraving. "I never thought," she said, "I'd be eating in the same room with Jefferson Davis. . . ."

In Praise of Practical Fertilizer

I raise my corn by horse manure and Vivaldi while my neighbor Mr. Collett raises his with commercial nitrogen and rock music. While I consider both horse manure and Vivaldi among the old, immutable things, the difference between Mr. Collett and me is the difference between several hundred acres and half a dozen rows twenty yards from the house.

I sit on the porch late, watching Mr. Collett's tractors cross the field in front of the old windmill, downwind of midnight, the music from his cab radio an indelicate fertilizer, yet I find it all good company. I look forward to planting time as though it were an occasion, which it is. The number of sensible rituals has waned in the world but planting, in any of its forms, seems to be among those left.

I am suspicious of many of modern agriculture's persistences but then I, moving placidly from garden to woodlot, live in the country without farming and therefore can entertain the luxury of suspicion. If I were a farmer by occupation, my suspicions would be sorely on trial. Character is what you are when hard pressed.

I like to listen to the old men of the neighborhood talk about their day, and I come away dazed at the recollections of their endless labor. This is not particularly work as a virtue, but work as fact. There was work man couldn't get around. There was work he couldn't get *over*, too. Contrary to kitchen samplers commending the virtue of hard work, there were men who died for their fields. A strong, intelligent man was equal to such work. Other men were less equal.

I like the stories of Jacob Spicer Leaming, the county's

preeminent corn farmer, and I regard him as one of the strong, intelligent ones. When Mr. Leaming went to the fields with his father in the first half of the nineteenth century, farmers used a bull plow, an unwieldy implement with wooden mold-board and a steel point. They furrowed the field, then without any other preparation, dropped the seeds and raked clods over them. "I have seen fields that looked like fields of boulders," said Mr. Leaming.

Sometimes a boy would come along and try to rake out some of the clods with a hoe but, as Mr. Leaming recalled, a third of his time was spent hoeing clods and two-thirds spent pitching the clods at blackbirds. Burs, Spanish needles, and pigweed grew eight feet high in the corn. Farmers said they could grab the pea vines at one corner of a five-acre lot and shake all the corn in it.

Mr. Leaming recalled a neighbor named Gray who used an ax to cut out the pigweed which, by harvest time, had grown two or three inches around. Mr. Gray pulled out some of the weeds by attaching them to his horse's tail. With the weeds gone, the corn fell over. At harvest time, a man took a horse and singletree up the rows, breaking the wild sweet potato vines, pea and man-of-the-earth vines, and knocking off burs so the huskers could get through. A common yield per acre was thirty-five bushels.

Another neighbor, Mr. Stiles, behind with his plowing, said, "I wish to God I had horses of steel with which I could pull a plow that would cut two feet and then I could go through and cover up every damned blackbird in the bottoms."

The steel horse would be some time in coming, but the steel plow arrived in this part of the country about 1830. Some of the farmers thought it was "a Yankee invention got up to swindle" them. But it was made of fine imported German steel and it rightfully did away with the old bull plow. The steel moldboard plow was not greatly improved upon in Mr. Leaming's time and hasn't changed much more since.

Mr. Leaming, his father, and two brothers took the steel plow, cut two or three inches deeper, cultivated well and kept out the weeds. "Then was I taught to let nothing green grow in the cornfield but the corn," said Mr. Leaming. In 1827, the Leamings averaged one hundred and four bushels an acre. The news, said Mr. Leaming, "went all over the bottoms and astonished everyone and was the talk and praise of all the landholders, and caused envy with many of the tenants. . . ." This year, I understand, the average yield in my county is predicted to be one hundred and eleven bushels an acre.

The Leamings did it with attentiveness and the use of what they called "practical fertilizers," the manure from barn and stable yards; and rotating corn with red clover which old Mr. Leaming recognized correctly as a builder of poor soil. "Many are running mad after bone dust, phosphates, and other manufactured fertilizers of various kinds," the younger Leaming noted, cautioning his neighbors.

In the autumn of 1855, Mr. Leaming was driving along the Bullskin Trace and found he had forgotten to bring feed for his horse. He stopped at a nearby field and asked the huskers to sell him some corn and liked it so well, he took along a bushel to experiment with as seed corn. In 1878, he won the silver medal at the Paris Exposition in Paris, France. Mr. Leaming had, as he put it, "beat the world's corn." Along the way, he had changed the growing practices of midwestern corn farmers. Even today, most of the yellow corn grown in the United States has been developed from the Leaming variety.

We tend to like stories such as this. We have a mythology of such stories that would compost the cornfields of Ohio. They are good stories and they appeal to a sensible, fundamental nostalgia in us for order and just deserts. There is nothing wrong with such stories, either, except for the fact that they often cause us to overlook the stories of *other* men. One man's meat is sometimes another man's portion.

It seems to me that farmers are more equal today. Machinery has the interesting capability of magnifying a man's strengths or his weaknesses. The problem therefore is not with the machinery but the notions we bring to it. We are a slow and opinionated species and have not yet come round to applying moral questions to the operation of our tools. But if I were a farmer, I would want to be able to consider the options of machinery and make what mistakes I might.

There was no particular virtue to the Leaming family's work. The virtue was in persevering. The options of the family were strength and intelligence and they used those qualities to overcome the rather stringent conditions of their time. The options of strength and intelligence, however, are quite often the options of luck. I admire luck but it's the cool, distant admiration I reserve for one kind of curiosity or another. For myself, I'd like to have it all, strength, intelligence—and a modest piece of equipment or two.

The Cold Boards of Morning

There is something about daybreak that takes even a man not normally inclined to metaphor and makes him fairly break out in it. A poetic fellow will, in the event of daybreak, lay about him with lyricism until there isn't a dry ear in the house. Daybreak is just one of nature's flashy little episodes that somehow has the power to consistently shove its observer off into pastoral effusions.

I am trying here to be an impartial observer, but let's lay out the biases: I am not a daybreak sort of person. I harbor the somewhat unpopular belief that life is more sunset than sunrise, and I have never been able to trust daybreak people. They make me nervous. I cannot seem to have any other reaction to anyone who leaps from bed at the behest of daybreak and comes down sprinting on the cold boards of morning.

This is not an appropriate gesture to start any day because it fails to take into account the full range of treachery that awaits in those interminable spaces between rising and lying back down again. The accurate response to beginning any day is to open one eye and take a careful reading before moving. Getting out of bed is an act of no little consequence, requiring both cunning and stamina.

When I was about ten, I saw a movie in which a diver crawled out of the ocean and suffered a severe case of the bends in front of a lady with a white bathing suit. In my young and untraveled mind, this distress had something to do with the lady in the white bathing suit. My father explained that the diver had risen too quickly from the ocean and the difference in pressure caused nitrogen bubbles in his bloodstream. The picture stayed in my mind, however, and ever since I have feared giddy ascensions and sudden surfacings. I have also stayed irrationally wary of ladies in white bathing suits.

Part of my antipathy for daybreak may be nothing more than a refusal to go along with an institution. Daybreak is one of our biggest institutions, in the category with such things as The Good Little Man, The First Snowfall, and A Service Station Man You Can Trust. What binds these seemingly disparate things together are our notions about them. Drop any one into a conversation sometime and watch the talk veer off into medium simper.

Daybreak got this way because America was founded by daybreak zealots. These people came here to worship daybreak, in any manner they saw fit, although it usually centered around one pagan practice or another involving early and violent exercise. Before long, the entire country had a genetic predisposition to standing in front of an east window at 6 A.M. As the country became more godless and people slept later, daybreak evolved into a *concept*.

There is, as of course there would have been, a whole aphoristic body of daybreak literature—you can imagine how those old Transcendentalist dogs worried such a meaty symbolic soupbone—but careful scrutiny reveals there isn't much heart in it. As for the non-Transcendentalists, my favorite is Tennyson, saying once at daybreak, "Come into the garden, Maud, for the black bat, night, has flown."

After going through the literature, I did an informal survey to see how folks were responding to daybreak today. A young fellow down the road told me, "I don't do a whole lot of daybreak," and one in town said that whenever he witnessed daybreak he knew it was time to be getting on home. He said catching daybreak that way made him feel a little guilty, so he tried to time things so he got to bed just *before* daybreak. "I feel fine," he said, "as long as it's still dark outside. As long as it's still dark, I've made it. I didn't stay out all night."

The Squire said he had been a sunset man for quite some time but he was getting back to daybreak, especially as he became older. He said he liked daybreak because it was punctual, it held up well, and was largely successful in resisting government regulation although he saw Daylight Savings Time as a dangerous trend, the height of political arrogance.

"Tampering with daylight," he mused, "seems to be wide-spread these days. . . ."

Another fellow told me about an experience he had with daybreak the morning his daughter was born. His wife went into labor about midnight so he rushed her to a hospital in Cincinnati overlooking the interstate. The child was born at daybreak, and he walked out of the delivery room and stood in front of the windows. "There was this river of lights from the cars," he said, "and in a few minutes the sun came up over the hills and illuminated the smog. My daughter weighed five pounds and six ounces. She didn't seem capable of taking on interstates and smog. I was also upset because here were all these people and not one of them recognized my daughter had been born. 'People have no sense of occasion,' I thought. I wanted a sign on the interstate that said, 'A child is born. . . .' Then, as the sun rose higher and burned away the murk, and after I watched awhile, it didn't seem like such a bad place ofter all. . . ." He reports that in the interval since, his daughter has grown rapidly and seems more like she might one day be able to take on the interstate.

A country singer I know, who has an odd turn of mind, said she had seen daybreak once and had been impressed with it. "I saw a family of Angus browsing across a pasture in the springtime, and the sun coming up behind them. I felt at one with nature. Or with the Angus. I've forgotten which, but I took some note of it at the time."

She said her experience with daybreak was that it was unpredictable. "It don't always do what it's supposed to do," she said. "You just never know what's out there. Sometimes a bunch of us get together and bang pots together, or wear false faces with dreadful countenances. Wasn't that what the old Druids did? Well, sometimes you just have to bully daybreak.

Of course, in the winter, it's so thin, it ain't hardly worth it.

"You know, the real early kind of daylight seems thinnest to me. I've often noticed it's thinner depending on how thick the night was. Daylight towards the first of the week tends to be a bit thinner, too. I think thicker daylight is better. Thin daylight is like thin air. Of course, that's just my preference. . . ."

I am not particularly proud of this fact, but in my own family, there is an old legacy of daybreak. My grandfather on my father's side arose in the dark, arrived in the fields before light and dozed on the plow handles until it was light enough to work. It was a habit he was never disciplined enough to break. When he was well into his seventies, he was still getting up before daybreak, sitting around on imaginery plowshares, and looking for some new ground to work. It was always an irritation to my grandmother. She was fond of saying that the last time he stayed all night with her was in 1948 when he came down with pneumonia.

My grandfather on my mother's side provided the genetic balance for this aberration. He was an old dairy farmer who loved his milk cows but disliked the dairyman's deviant hours. When he retired, he set his clock for 4:30 A.M. He did this every night before he went to bed. When the alarm went off the next morning, he shut it off, and enjoyed a moment of extreme bliss before turning over and sleeping another three hours. I have doubtless received my bloodline from this side of the family.

Once, returning from a trip after driving all night, I passed by the restaurant in the next village over. The sun was just coming up and the early indications were that it would be a flawless spring day, the kind a friend of mine said was reserved for the praise of poets and the forgiveness of fools. As I passed, an old farmer came out of the restaurant, fueled by a bait of Mrs. Barkley's good biscuits no doubt, looked at the sky, then across the freshly plowed Little Miami bottomlands, kicked up his heels and yelled, "YAAAAAA-HOOOOO!" Then he climbed into his old pickup and roared off into the sunrise.

I found that a compelling argument at the time, and even as prejudiced as I am on the subject, such articulation did give me pause. Perhaps the only thing wrong with daybreak is that it comes at such an unconscionable hour.

Mr. Baker Fusses into Summer

In early July of 1926, Mr. Baker drove into town where he paid a dollar fifty for a pair of overshoes. This was a prescient act on Mr. Baker's part, although at the time he was complaining of the dust.

"If the cracks in the cornfields lay the right way," he grumbled into his diary, "I could put in drain tile without touching a shovel."

He continued to plow, and to swear at the heat. "Hot enough to make Satan sweat," he wrote, describing the sky as being like a sheet of copper over his head. "Dry as Tophet," he said. "I should trade this horse for a camel."

It was ninety-five degrees on July 21. Mr. Baker finished plowing corn that day, noting that it had good color but not for long. It was dry for another week, then it rained. It was one of the biggest rains of the year. It rained all night on the twenty-seventh and didn't stop until noon.

Water was standing in the cornfields, so Mr. Baker waded out to pull weeds. This time, he swore at the rank-smelling

Jimson weed. "I need a crosscut saw to get some of it down," he said.

There was still plenty of moisture around by the second week of August. "Lightning in all directions this evening," Mr. Baker wrote. "This is the season of screech owls, smelly cornfields, noisy cicadas, and hot muggy nights. It is great growing weather."

On August 19, it rained another three inches. The creek rose the highest it had been in thirteen years. On August 20, Mr. Baker noted, "It rained last night. It rained this morning. It rained this afternoon. I have no reason to believe it won't rain tonight." The garden turned black and began to rot.

On the twenty-first, more rain. "This on top of a world already sodden and rotten," said Mr. Baker. "Vegetation is rotting and disaster threatens. To make matters worse, an old sow died last night (she may have drowned), willing me eight piglets to be raised by hand. I spent the afternoon struggling at the hog business. Distress and loss."

The weather cleared on the twenty-third and Mr. Baker made cider. His manner improved. The next day he went out to mow but the horseflies forced him to quit. "The horseflies were so thick and bloodthirsty," he wrote, "I considered returning to the house for the shotgun. Worst I ever saw. Must have come in last night when I left the orchard gate open."

He dug potatoes on the thirtieth, pronouncing them "poor." He also cut weeds. "Corn prospects poor," he wrote. "Weed prospects excellent." He noted that the heat was back and the pasture "exceedingly short."

By the first week in September, Mr. Baker was swearing at the heat again. "The sun," he wrote, "has fried half the world. . . ."

Passion among the Limas

Miss Mary sat on her porch overlooking the garden and told The Squire her theory of the lima bean. "You cannot *trust* limas," she said. The Squire was surprised. He was a confirmed lima man. He asked Miss Mary why not.

It was her turn to be surprised. She looked at him over her glasses, in the way a schoolteacher fixes a backsliding pupil. "They are," she said, "simply not trustworthy. If you had ever been intimately involved with limas, as I have, you'd know."

The Squire almost rose to that bait, but thought better of it. Instead, he circumspectly asked Miss Mary about her experiences with limas. "I don't know that I wish to talk about it," she said. She rocked a few moments without saying anything. So did The Squire.

"Oh, all right," she said finally. "The first time I was out on my own, I planted limas. I treated them like all the other vegetables. No favoritism. We were all in that garden together. I waited and waited. But no limas. I could *still* be waiting.

"I stayed up at night. I paced. I lost weight. I tried lecturing. 'Someday you'll understand when you have limas of your own,' I said. 'I have worked my fingers to the bone,' I said. Those limas let me down. But did I condemn them that year? No, I did not. Am I a heartless person? No, I am not. I gave them a second year, a third year. There were still no limas. . . ."

Miss Mary paused. She sipped her glass of cooking sherry. In the garden, a cabbage moth colonized the broccoli. Miss Mary frowned, which seemed to displace the moth, and it flew off.

"I will admit limas were an afterthought that second and third year," she said. "I was no longer a naïve, simple country girl after that. I had grown up. So I planted limas last. I gave them the same care, but perhaps I was somewhat distant. Maybe this is why they did what they did to me. Maybe it is personal. I tried to think if there was a family disposition toward this kind of failure. But then I recalled my grandfather had a splendid relationship with his limas.

"Green beans never let me down. Neither did the spinach. You can trust spinach although late in the summer it gets crotchety. Lettuce does this, too. Lettuce has a bad attitude late in the summer. Both spinach and lettuce are what I'd call your short-time nice vegetables.

"But you take the green bean. *There* is a vegetable. You can put money on the green bean. It is big and strong. You pick green beans and they grow some more. They are no fair-weather vegetable. They stay around until the frost then depart gracefully. What more could anyone ask for?"

Thinking of green beans made Miss Mary felicitous. She poured herself another glass of sherry. The Squire, not a drinking man, chewed thoughtfully on a clover stem. In the garden, the moth was in a holding pattern over the chinese cabbage.

"I like corn, too," said Miss Mary, "but I can't think it belongs in a garden. It belongs in a field. Corn is too *tall*. It intimidates the other vegetables. It leans over the cauliflower and makes deprecatory remarks. The cauliflower doesn't need to hear this kind of talk. It already *knows* it's short. I like melons, too, but I find them malcontents. They push and shove and won't stay put. I think they scare the vegetables."

Melons brought Miss Mary back to limas, although it was obvious she wanted to be elsewhere. "I think limas influence

the other beans," she said. "Limas make other beans . . . *dull*.
Limas incite other beans to sort of lie down. Limas are like
liver. No one wants to eat liver. No one wants to eat limas,
either. Do you hear folks say, 'Why don't you and Gladys
come over tonight and we'll have a few drinks and cook up
some limas?' No, you *don't* hear folks say that."

The Squire tried to recall if he had ever said that and, even
though a lima man, he knew that he had never invited anyone
over to cook up some limas. As a lima supporter, he still had
to acknowledge Miss Mary's argument.

"Limas are very unsexy, too," Miss Mary went on. "Now
for sex, you take your zucchini. The name sounds like an
Italian count. Zucchini is a vegetable with manners. It grows
in a nice little circle and doesn't bother anyone. Very courtly
vegetable, zucchini."

Miss Mary drank the rest of the sherry and tossed a rock at
the cabbage moth. "All in all, though," she sighed, "I guess
I'm just a green bean person. . . ."

Heat

The heat was another presence in those summers. It was like
a taciturn relative we accommodated. We moved over a bit,
and down one at the table, and that was pretty much that. We
didn't even complain very much. The Royal Theatre was the
only air-conditioned building any of us had ever seen so we
associated the technology solely with theaters.

It was a time just after The Bomb and just before every
possible breathing current in the country had been captured

and somehow sold. We didn't complain about the heat because we knew just enough to be mute in the presence of the unspeakable. A list of the unspeakable was considerably larger then. It contained such things as summer drought, electricity, and most anything north of the state line.

The heat began, earnestly, in April. That was when Leonard Stenhouse's momma let him out of the long johns she had sewn onto him the previous October. Leonard was the true harbinger of summer. When Leonard heated up along about April, school, for all practical purposes, was over. We left Leonard's ripe effluvium and went off into the rising heat of summer, to see what we could find.

There was heat everywhere: heat lightning, heat mirages, sweltering animals, tin roofs that cracked and banged. My cousin Pressley, bored insensible in the deeps of July, fried an egg on the hood of his dad's Buick. Then he produced a fork and ate it off the car. He pronounced it "passable." Down the road, Aunt Soonie had a setting hen she wanted to break. She chased the hen away by pouring water on it. When she left, the wet chicken returned. Uncle Frank saw it on his way in to dinner.

"Soonie," he said, "it is hotter than it has ever been before."

"Why do you say that, Frank?" Soonie asked.

"Well," Frank said, "I saw your old hen a few minutes ago and it was sitting out there just *a-sweating*."

Church seemed the hottest place. Through the open windows, I watched heat waves shimmer along the tombstones and in front of me, insects droned around Sonny Boy Verdin's hair-waxed head. The minister loved the summer. In the winter, the cold made him squeak but in the heat he expanded, converted himself to a gaseous form, and hovered volatile over us. I hated it when he came home with us for Sunday dinner. There was piety in the air, which tempered the fragrance of chicken. On those Sundays, the gravy seemed thinner.

I worked out my guilt in the hayfields. The hay mow was purgatory, the temporary punishment that purified. It was one hundred twenty degrees under the tin roof. If the minister had ever said hell was an eternity of Julys in a hay mow with a tin roof I would never have argued with my mother nor tried on various occasions to look up Mary Frances Verdin's dress. I would probably not be a sinner today. The minister's exhortations about brimstone left me unmoved. What did I know of brimstone? Brimstone, as far as I knew, was mined in some far off place and exported. But hay mows I knew about.

Today I fear many things, most of them unknown. I still do not know what brimstone is, but I worry about it. In those days, I feared only what I already knew about. To my untraveled frame of reference, the hay mow was one of the world's fearsome things. And while I feared it, I liked it, too, a fact curious to myself even at the time. I liked it because such work was expiation; it was atonement for every heretical thing I had done or thought over the winter, most of it known only to me. After days in those sun-blasted fields, I was innocent again. I could go forth and sin anew. This may have been my first intellectual act: I had, my own self, fused Baptist and Catholic theologies in one embraceable faith.

The other reason I liked the fearsomeness of fields and mow was the extremes they offered. Shade was never so blessed, nor plain water so sweet. At night, as the heat slipped slightly its hold on the earth, I slept the untroubled sleep of the newly born. And on days when we finished early enough, my father sent us off to Lake Taro, the big spring-fed, rock wall-enclosed swimming hole, a couple of miles away. In the aftereffect of heat and field, Lake Taro was so pleasurable as to stop the heart. I was purely mindlessly happy, like an otter or some other water-creature in its element. I have rarely felt that way since, an instruction. Perhaps the truth lies in extremes. As careful creatures, extremes make us nervous. We try to avoid them. Yet in extremes, we are most alive. Consider the heat of *that*.

At the end of July, Lake Taro closed for a week. A great covey of Baptists (more conservative, even, than my own) lit next door for a week-long revival. They flushed at every sin, mortal or venial, drumming the heavy air. Because both sexes in the same swimming hole was high on these Baptists' list of utter proscriptions, there was no swimming for a week. The temperature of the surrounding countryside rose another ten degrees.

I cannot remember that we complained much, even then. The Baptists were surely a knot in the drawstring of life but they taught us the meaning of words such as "discrepancy," and gave us a sense of humor. To be fair, they got some of us, too. My own cousin Pressley, after a promising start, would one day marry a religious music major who said such things as "I never say 'Gee' because it's too close to 'Jesus.' "

During revival week, we hung around the icehouse more than usual. The icehouse was the scene of one of the neighborhood rites of passage (the hay mow was another). It was run by old Mr. Ledford who, at some moment known only to himself, would appraise each youthful frame and pronounce it mature enough to take down the tongs and wrestle a hundred-pound block of ice out to the loading platform. The walk into the ice room was delicious. I have forgotten the aisle I walked down to graduate from high school, but I have never forgotten the sawdust aisle in the icehouse.

We got mean by August. Putting up corn, Pressley and I bribed his little brother Albert into standing still while we glued cornsilk under his arms. That weekend, we filled Lake Taro with stolen watermelons. Aunt Bert said it was the heat.

Cool weather came, town jobs, graduation. I crossed the state line, fell in with a faster crowd. I went to some big cities and heard me some big talk. I met some Episcopalians.

But in the summers, when the heat does not fall with the dimming of the light, I ponder the wide swings of temper and desire. I leave the air conditioning and sit out on the porch, feeling the heat. I recognize then I have never wanted the temperate climate.

Grazing Rites

There are several things I like about the county fair but I think the main thing I like is that it gives city people a chance to step in cow manure. This is always an humbling experience, serving in a small way to remind a man that each footstep of his life is apt to be more precarious than he thinks. It is certainly an experience that should not be denied one simply because he is not fortunate enough to have his own supply, and it still warm with the manufacturer's imprimatur.

The county fair is also one enterprise that is packaged pretty much like it always has been. The barns are still full of real animals and, thankfully, promoters have not yet replaced the old skills of familiarity and patient grooming with electronic sheep being maneuvered across unsullied astroturf for points.

The Squire and I watched the sheep-judging first, then we headed over to the cattle barn. I consider The Squire a qualified fair-guide because he has raised everything from Angus to an opera singer and as far as I've seen, they've all appeared to be out of good stock, although a few eccentric qualities crop up here and there.

We passed by the fowl barn, which reminded me of Mrs. McDermott, who used to buy eggs from The Squire but only once a year, late in the summer. The Squire found this curious but never said anything. Then he ran into Mrs. McDermott at a poultry exhibit one fair-time and she pulled him over to one of the exhibits and said, "Do you recognize these?"

"Well," said The Squire, "they look like handsome eggs, and that looks like a blue ribbon on them."

"Those are your eggs," she told The Squire, "and I've won that ribbon for several years now."

The Squire got out of the egg business shortly afterward, but he enjoys the story himself.

The dairy cattle judge was listing the qualifications of a big Holstein when we walked up. "She's making up a real fine udder there," said the judge. "I admire her front end although I'd like to strengthen her loin a bit. An exceptional set of feet and legs." I always enjoy the dairy judge's delivery and it never fails to remind me of beauty contests. Dairymen would make fine judges for the Miss America contest, although out of habit they'd likely encourage the well-fed contestants.

Like The Squire, I'm an admirer of the cow. She is not as smart as the hog and she has the somber regularity of an old Methodist, but she is a vast repository of serenity in the troubled pastures of her existence. She is humble and perseverant through pain or bad weather and always seems to suggest she knows what is what.

It is a highly sensible measure for nature to have provided the cow with a four-chambered stomach where the good beast,

at some leisurely later time, might recall the earlier grandeur
of her dinner and digest it contemplatively, at peace with her
surroundings.

Before we left for the day, The Squire suggested that it was
sometimes difficult to tell who was on which side of the fence,
and exactly who was watching whom. I think that's right. The
human animal is an egocentric creature, certain through each
of his poor turns in the ring. A Holstein, for instance—even a
registered one—would not have cared much about conquering
Gaul, settling instead for merely grazing it.

Mildew

Ah, this steamy, moldering month. One must check everywhere for signs of mildew and rot. The dung beetle is all
around, watching for us to remain motionless a second too
long. He's a princely insect in August, finding at last his dank
province. He's back in power after exile. Crews work around
the clock to keep various kinds of ivy from covering Clarks-
ville. Strange mushrooms grow overnight and in the morning
we may stand upon them to catch a glimpse of the retreating
horizon. The piano sweats. Uncle Walker has been overtaken
by kudzu. I stand before the mirror daily and check my tongue.
Strange growth everywhere in this fertile air. I carry a machete
to hack my way through overgrown thickets of words. Even
worthy constructions might at a moment's notice be covered
over in crabgrass and ragweed. The times (such times!) demand
a bush-hog, native bearers with honed scythes in swingtime,
bodies bent on weeding out discrepancies in the late gardens
of our intent. Nothing short of a war effort is needed. We

seem, however, to have grown full and fat. Are we up to battle? Our sentences grow sluggish, fueled by intellectual gravy.
The belt of our instinct which holds up, thusly, the trousers of
our action, gains a notch. Politely, we speak of deeds while
eyeing the sitting room for a likely divan. The mind reminds
us constantly that it is Sunday afternoon, August, and waltz-
time. Smells hang in the air so tangible as to be gathered in
bushel baskets and dried out, hung in the cellar for winter
when there's a dearth of smells and the nose cries out for
amusement. Piglots. Creek bottoms. Graves. I have seen fish
roam with impunity through the bottomlands. What do we
care, momentarily, if our artless constructions cave into the
cellar, and the orchards of our speech fall bruised and spotted
around us? We grunt our will at one another. Shrug, tilt, bellow, rasp, and thump our poor meanings out, housewives
seeking a proper melon. We have never been very successful
with language, anyway. Perhaps we need to begin again. Perhaps: A language based upon the exhibition of fresh vegetables. To speak, one must garden, else maintain a strict silence.

Requiem for a Hotel

I was sorry when the old Xenia Hotel was not rebuilt after the
tornado. I am sorry every time I drive past its empty lot, for
myself and for my appetite, which still rises up unchecked, as
though momentarily the two of us would be sitting down to
pan-fried steak and a salad followed by a piece of homemade
coconut cream pie.

A friend of mine finds an amount of irony in a Providence
that, as he puts it, hits something like the Xenia Hotel, yet

misses the Armory. The old place was an oasis in a desert of franchises and if the food was simple fare, which it was, then the diner could finish off by drinking in the aroma of the place.

The dining room was large and high-ceilinged and the diners sat at small tables with starched tablecloths, on straight-backed chairs that might have come from around someone's kitchen table. The room was very quiet, and the waitress always brought out butter and a small loaf of freshly-baked bread on a wooden board as soon as she had taken your order.

When dinner was finished, the diner walked out into the lobby and paid the clerk behind the hotel desk. The lobby was filled with antiques, including a wind-up phonograph, an apple peeler, and twenty-five rocking chairs. Usually, a half-dozen or so old retainers were ensconced in the rockers, alternately dozing and watching the traffic pass.

Miss Mary Dakin could usually be found in one of the rockers. She had owned the place since her father, Jide, died in 1934. His desk was in the lobby, beside the apple peeler, and so were the family chests. As for why these things were in the lobby, Miss Mary said it was for the same reason that the dining-room chairs did not match.

Miss Mary grew up in the hotel. There were fifty rooms, and the family slept wherever each member wanted to, although Jide preferred the third floor. He said it was health-ier. "I never know where your mother is in the morning," Jide sometimes complained to Miss Mary. Then he went down-stairs and fixed pancakes and cabbage. Or whatever else he happened to think of to go with pancakes. He also liked pie with gravy on it.

He was up at four, but he slept in the rockers during the day. Sometimes, Miss Mary said, he and a whole contingent of acquaintances would be there in the rockers, sleeping in concert.

Jide told Miss Mary everybody should be off the streets by nine in the evening. "Nobody *decent* out at such an hour," he told her. He also said he didn't know why anyone wanted to travel. He was once a buggy salesman and had been everywhere, and could tell anyone else what was there.

Miss Mary's sister ran away and married Bert Black. "I think he is a good man, Bert Black," Miss Mary said. "Blacks are black!" said Jide. "And Longs are long," said Miss Mary because, as she explained it, "we had a neighbor Long who was short and three feet wide." Her logic was found wanting. She was chased out of the room.

Miss Mary once helped at a wedding breakfast for a young couple from Antioch who wanted eggs and asparagus. They told Miss Mary that green and yellow was their color scheme. A neighbor lady, who was very fat and couldn't see over her stomach, was continually getting stuck in the rockers. She told Miss Mary she cut her toenails on faith. She said Miss Mary was the ugliest kid she'd ever seen. This did not bother Miss Mary because, while she did not know for sure, she suspected she was quite pretty.

Miss Mary took singing lessons but she didn't want to sing. She was too timid. There were recitals and Miss Mary, who had a good voice, hid.

She went to Wilmington College for awhile and the dean was once asked who in the room was most likely to succeed. The dean said, "Well, Mary is *least* likely. She doesn't put her mind to things." Miss Mary said that was true. "I took piano lessons and stared out the window," she said. "You can't do too much with people like that."

For a year, Miss Mary played the piano in a nightclub in Mansfield. Once, a drunk walked by and said, "Do you *have* to do that?" Miss Mary said it was no life for a timid person. So she came home to the hotel and stayed there.

She died a few months before the tornado came, almost as if prescience was a contributing factor. I don't know what they'll build on Miss Mary's vacant lot, but I doubt if it'll be a hotel with a lobby full of rocking chairs. I expect we've seen the last of the coconut cream pie, too.

Baseball on the Backlots

Hunky Haley always thought of baseball as a country sport. It was of little matter to him that baseball was played mostly in the big cities like Cleveland and Cincinnati, under electronic scoreboards that set off nuclear explosions when the home-team scored. In the hinterlands of Hunky's mind, baseball was played in the cow pastures and back lots, just as it always had.

As a boy, he watched the villagers play on the big skinned diamond behind Bernie McKay's house. Hunky had to sneak out of the house to watch because Mr. Haley didn't want him up there. The men drank beer on the long, sultry Sunday afternoons and Mr. Haley disapproved of beer. Hunky envied Pigeye Patterson because *his* dad drank, and let Pigeye play outfield. Hunky also envied Pigeye because Pigeye could swear better.

Most of the players were grown men, but Pigeye was big for his age. So was Hunky, for that matter. That was why he was called Hunky. He could have played, too, except for the beer. His father's edict made him decide that as soon as he finished school, he would go away somewhere where he could play baseball and drink beer for the rest of his life.

Everybody played on the village team. If you could walk, you played infield. A cripple went to right field. If someone was missing from around the house, he was likely at Bernie's diamond. Once, Susie Haydock asked a little boy in the street, "Do you know where Jim is?" And the boy said, "Yes'm. He's up to bat."

Hunky's idol in those days was Rube Robinson, who lived in the village and played with the semi-pro teams around Dayton. He had offers to play professional ball but he didn't want to leave home so he said no. The sportswriters called him "the player with the million-dollar arm and the ten-cent head." Rube didn't seem to mind.

Once, Mr. Patterson took Hunky and Pigeye to see a Saturday afternoon doubleheader in Dayton and Rube Robinson pitched both games. He said the effort was due to a quart of corn whisky. He drank half of it before the first game, and rubbed the rest of it on his arm before the second.

Rube Robinson's two best pitches were a change-up and a knuckleball. His fans called his knuckleball "the barleycorn fadeaway." The knuckler, he said, "was a sight on this earth. You'd know what I mean if you ever tried to hit a hummin' bird with a broom handle. It just went up near the platter, stopped, danced a little jig, then exploded. I never knowed whether it was gonna break inside or outside. I was known to have needle-threadin' control. It started when I was a kid throwin' acorns through knotholes in the side of the barn. Worse whippin' I ever got was once when my old man handed me three smooth rocks, sent me out in the woods back of the house, and I come back with only two squirrels. . . ."

Listening to Rube made Hunky love baseball talk. On certain intense spring afternoons when the air was still cool but the soft maple had budded, he and Pigeye would go into the county seat and watch the college play. Few people came, so the two of them sat in the bleachers and analyzed the games:

"We're down 5-2 in the ninth, one foot in the grave. Bailey walks, Chapin singles, Hewitt gets the free trip to first. Three ducks on the pond. I call out to their backstop, What's the matter, boy? Lost it in the sun? Johnson bangs a low liner that hits the shortstop, Johnson down. Everything's still swell. Sacks jammed, Shook up. Where's Shook? Shook! Daydreaming in the dugout. DAY-UM. Shook comes off the bench, BLAM! Double! Two runs across, it's 5-4, down by one, Miller swings, sacrifice to right, score's tied, Carter walks. Get Shook home, we're on Wall Street. . . ."

Hunky's earliest ambition was to be a catcher, or an Indian. Mrs. Haley remembered that on Hunky's seventh birthday he asked for an athletic supporter because he thought it would make him run faster. Hunky thought a lot about being an Indian, too, because he liked the idea of being outside of everything. This idea was something like running away and joining the circus. Hunky thought about running away and joining the Indians.

He liked all the tribes he read about except the Pueblo. The Pueblo were town-dwelling farmers, and it didn't appeal to Hunky to plant corn because that was what Mr. Haley did. If he had studied his ambitions more closely, Hunky perhaps could have consolidated them by becoming a catcher for the Cleveland Indians, a team perennially outside of everything.

Hunky wanted to be a catcher because catcher was an indispensable position, like quarterback, but without the attendant glamour. Hunky was a shy boy, and wary of glamour. He liked the sense of harmony in baseball. He liked blending in.

His senior year in high school, Hunky did not even go out for baseball. He was a good baseball player—he was a catcher—but he liked *watching* baseball better than playing it. His problem was that he was not *competitive*. Conflict was upsetting to Hunky. His sympathies changed in the middle of a game, even ones he was playing in. He remembered playing a Little League game and after his team won, the other team burst into tears. Hunky felt terrible. In baseball, a .333 batting average meant excellence. To Hunky, it meant the batter struck out two of every three times. Not playing baseball his senior year made Hunky somewhat unpopular.

He went to the county seat college where he studied to become a teacher. He continued to watch baseball, not particularly as it was played in the big cities, although he seemed to know the most recondite baseball facts.

In the proper months, when the heat rose in the afternoon and the light seemed exactly right, Hunky (by this time, of course, he was no longer called Hunky) walked down to where the college was playing. He liked the slowness of baseball, the leisurely pace. He recalled Mark Twain writing that baseball was a *modern* sport that caught the nervous, driving energy of industry.

Baseball to him, however, was open fields, and heat. Baseball allowed Hunky to look backward over the century to its beginning, and forward to the remainder of it. It was a sport that generations passed along, like furniture or predispositions. In his mind, Whitey Lockman and Walker Cooper existed simultaneously with Pete Rose.

In the sixties, in that decade of failure and acrimony when it seemed all the college campuses in the country might burn down of merely spontaneous combustion if not outright arson, Hunky thought of leaving the country. But in one of the worst convulsions of that time, he found himself following a pennant race.

This made him feel better. Not the game itself, but the

recognition that he was hopelessly *American*. At a faculty meeting, he found himself saying, not without an air of amused detachment, "Let's put aside racism and pollution for a moment and talk about something *really* important: *baseball. . . .*"

He continued to go down to the college field, and later in the season, he watched the town teams play. He did this all his life. It didn't particularly matter who the teams were. He may not have ever known. What mattered were the long, hot afternoons, the endless afternoons, the smell of fresh cut grass, and the sense of leisure and harmony.

Hunky was past seventy when he died. It was in the fall of the year Cincinnati won the World Series. That was how Martha, his wife, remembered when Hunky died. She had always been the *real* baseball fan in the house.

Ruminations on Cow Manure

There were not many contestants for the cow-chip throwing contest at the county fair this year and nobody, participant and spectator alike, seemed to be having much fun. It is a curious and contradictory event when sober-minded people throw cow manure. Cow manure throwing is just not for the sober-minded. A person of such a set of mind should be at home, doing accounting or brawling with the crabgrass.

Mr. Anderson might have been the only one having fun. He not only entered the contest and got off a respectable throw, but he was responsible for the wagonload of ammunition (the wagon was also entered in the parade, driven through town accompanied by a cow-chip queen and two cow-chip princesses). A man has to have a sense of humor to allow him to wander around in a pasture sorting out cow manure and stacking it on a wagon.

Mr. Anderson, a pleasant man in overalls and a baseball cap, said that such a pursuit can make a man feel a little silly. "People are driving by out on the road and everytime someone comes by you pretend you're just out for a walk in the pasture," he said. "Once I was working away and I got the feeling I was being watched. I turned around and there was a cow standing right behind me, like, 'Where you going with that?' Well, I done it the first year, you know, and it just kinda stayed on me."

Mr. Anderson's greatest moment came a year or so back when it rained every day before the fair and, in the best of both pioneer spirit and Yankee ingenuity, he collected a wagonload of wet chips and dried them all by baking them on his backyard grill.

There is, too, a proper technique involved in this collecting. A man named Dick Workman explained this technique to me several years back when he organized the first contest. "We were a bit naïve going into this," said Mr. Workman. "It was trial and error. We got most of the ammunition from Ed Michener and Ernie Cook and most of it was from cattle fed on grain. Well, that produces an inferior end product. For the proper density, you need pasture-fed cattle."

Mr. Workman was a good organizer, however. At the entry booth I recall that he had on exhibit a flawless cow chip, of masterful design. In the middle of the chip was a perfect, newly-sprouted corn stalk. It was enclosed in one of those large clear plastic cake dishes and it was quite handsome. "There is definitely an art to throwing one of these things," he told the contestants. "Unfortunately, I don't know what it is."

It was quite a contest that year, and there was a lusty crowd. When the mayor walked into the ring to toss out the first chip, the crowd booed and made him take off his gloves. Later, one of the contestants heaved a large chip which broke into several pieces, part of it scattering the judges. "Look at that," said the announcer. "Still got the handle in his hand."

After the dust had settled and a barber named Van Nuys had won the mens' division (there was also a division for politicians) with a rather spectacular toss of 157 feet, five and a half inches—at that time less than 5 feet under the national record—Mr. Workman stood on the bare, chalk-lined field which resembled some Armageddon of the digestive tract and said, "Well, all the winners were local boys which just goes to show you, well, I don't rightly know *what* that goes to show you."

There is not a lot of literature on the origins of such a sport as this, although I suspect it began when one or another group of high-spirited pioneers, picking up buffalo chips for making a fire, began trying to see who could sink a chip into the wagon from the fartherest point of the North Forty.

When I was young, the boys of the neighborhood organized some great battles with cow pies. Like Mr. Anderson, we sought the perfect chip. In this case, the golden mean was a chip baked on the outside by the sun but still soft on the inside. It was ideal to throw, with devastating effect. I suspect that the modern-day equivalent of this is the frisbee, which seems to me nothing more than a plastic cow pie with all the mystery taken out of it.

Like technology rising to grander stations, we graduated to oil cans filled with sand and shot them across the neighborhood by means of an old inner tube cut in two and tied on either side of an upstairs barn door. This escalation in the neighborhood arms race ended when Presley Melton shot an oil can through his uncle's dairy barn window, causing a large Holstein named Valentine to jump up and down on the Melton's new DeLavel milking machine. Mr. Melton then jumped up and down on Presley, and as Lewis Mumford might have said, our technological aspirations were profoundly altered for a time after that.

I am disturbed by the atmosphere at the fair, though. It seems to me that people just do not enjoy throwing cow manure as much as they used to. If this is a fact, as I suspect it is, then I am sorry. Of all the things man chooses to toss around in his idle moments, from baseballs to quoits in the country, the cow chip seems to state its case the best. A man throwing cow manure knows exactly what he has got hold of, and there is no danger anyone will make anything more of it.

Country Correspondence

A man named Fleischman and I have taken to corresponding from time to time and since I live in a midwest cornfield and he in Los Angeles, these scribblings provide a fine release for the notions each of us harbor about the other's place of residence. These letters run like overland salmon, spawning prejudices and pleasing both of us.

I hold to the notion that the new world is being born in Los Angeles, and from all the noise, it sounds like a breech baby. Mr. Fleischman worries that I may become active in the Masonic Lodge.

When I was writing a book about the village of New Burlington, I received a letter from a Mr. G. A. Feeney, postal inspector. The return address was curiously the same as Mr. Fleischman's. In his letter, Mr. Feeney said he was investigating reports that fragments of this village had been gathered up by a local author and mailed off to New York. If so, he said, then this was entirely improper since the village was government property. Mr. Feeney did admit, in his letter, that there were other rumors, to wit, the village had been borne off to heaven by an angel of The Lord driving a flaming manure spreader.

When the book was finally published, Mr. Fleischman wrote to say that it, when opened at his desk, exuded a hearty fragrance compounded of near equal parts of "river mud, coal oil, manure, rust, the syrup from the bottom of the silos, and the smoke of an old man's pipe rising in placid defiance of the Methodists." He became fond of referring to it as "that widely misunderstood book about the persistence of chicken thieves."

For a time, he was given to sending off descriptions of both his country and what he remembered of mine (Mr. Fleischman is at an advantage here because he is a well-traveled man and has seen our cornfields while I refuse, upon well-founded provincial intolerances, to visit California). "The ground here," he wrote of Los Angeles, "is all pale with browns and yellows. It is so rich in light yet so niggardly in substance, and all the wild places look unfinished as if only half the materials have arrived yet."

He once described a summer afternoon in Ohio this way: "Humidity so high the mosquitoes are sticking together. Cows bra-less in the withering fields. The average pulse rate on the old mens' bench at the courthouse drops another beat. . . ."

After I told him I had spotted a Mercedes in the neighborhood, he responded to my distress. "I am sorry to hear a Mercedes has moved in, bringing all the lawyerly riff-raff that seems to go with it," he wrote. "First thing you know, you'll see a 240SL up on blocks in the weeds. Personally, I've got nothing against a Mercedes but they do smell funny. And, sure, you can haul syrup water in a Mercedes but did you ever try to plow with one? I put a pair of John Deere high side wheels on one but the damn contraption tipped over on its face before I did more than an acre. And the price of a good collar for a Mercedes is frightening. They also compact the earth something fierce. . . ."

Once, Mr. Fleischman wrote and outlined a company he was thinking of founding. The company would manufacture something called a "High Rise Hog Farm Kit" for people who had an extra room in their apartment and were eager to earn income at home in their spare time. Mr. Fleischman even had a tax write-off for using the bathroom as a set-aside.

In January, when Ohio was up to its scuppers in snow, Mr. Fleischman wrote to say he was organizing a privately formed Snowmobile Marathon for Christ which had agreed to carry in needed supplies. He said times in Los Angeles were hard, too. "We ate the horses last week and I'm considering selling Bertha and the kids. Fella offered me $15 for the set. Bertha's kinda sulky anyway. . . ."

Mr. Fleischman's finest moment came, however, when he wrote to tell me of a dream he had, in which he saw the children of Cincinnati flocking to Riverfront Stadium to plant potato patches. Suddenly, I began to see what the term "farm team" was all about, and ever since, I have had a persistent image of sixty thousand people cheering an acre of tomatoes ripening in the infield.

Last Respects for Mr. Humphrey

Johnny Rhubart was a great walker. He thought nothing of walking to Xenia which was eight miles up the road. Sometimes, on his way to Xenia, he met Jonathon Ellis driving his coal wagon. This was what gave Ike Peterson his idea.

The next time Ike was in the barber shop, he proposed a race between Johnny Rhubart and Jonathon Ellis's coal horse. The loafer's bench worked on Mr. Rhubart and Mr. Ellis, and the race was on.

They started in front of the barber shop, and ended in Xenia under the courthouse clock. Johnny Rhubart was there ten minutes before Mr. Ellis's coal horse. Mr. Ellis took a lot of kidding about it. "Jonathon," said Luther Humphrey, "have you considered hitching Mr. Rhubart to your coal wagon?" Mr. Ellis pitched a lump of coal at Luther.

The fellows at the barber shop thought it was likely Ike Peterson set up the race just to see what Luther would have to say. It seemed Luther always had the last word.

Once, Ike and Luther were taking a load of corn to the livery stable in Xenia and, coming back, Bascom Hartsock's bull was standing in the road. The bull ran across the road, Ike's horses shied, and when everything stopped, the bull and two horses were under Ike's wagon.

"Hello, Ike," Luther said, looking precariously down off the wagon. "Wouldn't that work better if your horses pulled the wagon rather than carried it?"

Ike was a Republican and Luther was a Democrat but they seldom talked politics. When somebody stole Ike's hams out of the smokehouse, Luther came by and noticed only one was left. "Whoever did it must have been a Democrat," Luther said to Ike.

"How you figure that?" asked Ike.

"A Republican would have taken them all," answered Luther.

When the loafer's bench recounted Luther's various last words, the one that came up most frequently, however, was the one involving Luther's stint as an undertaker's assistant.

Luther had been hired by the undertaker to dig up the remains of a young reprobate and ship them back to the wealthy family back east. Later, Luther confessed that what went back east weren't the remains of the young reprobate but those of old Lamar Collett.

"Well," he explained to the fellows at the barber shop, "I was sitting out there on Lamar's marker before getting started, and I thought here was this easterner who had been all over the place and done everything, and beside him was old Lamar, who had never been out of the county and whose main ambition was to take a train ride. So I said to Lamar, 'Lamar, here's your chance. I'm gonna give you that train ride. . . .' "

Luther was a baseball fan, and before he died, he told Ike he wanted a baseball game played at his funeral. Ike made the arrangements, and played centerfield. The fellows played six innings, then went over to the grave and paid their last respects to Luther. This time, Ike had the last word.

"*Safe*," he said softly, as they lowered Luther's box into the ground.

On Storytelling

I have never liked jokes. I have a severe allergic reaction whenever someone sidles up to me and says, as joke-tellers invariably do, "This is the funniest story I ever heard." The
assumption here, as wrong as most assumptions, is that humor
is a common language and we're all in it together, like English.
In fact, humor is like those old European duchies with everyone squabbling in several tongues over disputed borders.

I have two reactions when someone presses a joke upon
me. If I'm feeling at the height of my powers, I try to bear up.
If I'm not, I shoot the perpetuator and try to make it anything
but a flesh wound. A person carrying around a joke, however,
is like Typhoid Mary; sooner or later, he's bound to give it to
someone. (I say "he" with generalized impunity, for it seems
to me most women have the innate good taste *not* to tell jokes.)

A joke is always announced, too, as though it were an
important candidacy, or an incoming flight. This is the awful
part of joke-telling because it lets you know what you're in for,
which is laughter, whether you want it or not. You know you're
going to have to come up with some laughter for, after all,
that's what a joke is about. Someone comes up to you on the
pretext of *giving* you something, then presents you with a due
bill, a curious state of affairs.

I know joke-telling has probably been around as long as
language has but it still strikes me that the form is very modern. A joke is nothing more than pre-packaged humor. For
instance, a man dumps a joke in the pan of sociality, adds the
water of his own enthusiasm, and voilà, he has an entree for
which he gets all the credit. This is finally the language as

ready-made pizza or pie mix and it rarely fails to give me indigestion.

I'm very partial to storytelling, however, although I admit I have my own definition of what makes up a story. While a story *may* be borrowed from someplace, as a joke most always is, the story more naturally is something within the teller's vision, something he's seen which struck him and he'd like to pass along. It has a narrative line, too (this differentiates it from gossip), and seems to get better with retellings which tend to add texture, like a kettle of soup picking up fragrance in the reheating.

A good piece of storytelling should have a good first line, such as this: "It was noon when they threw me off the hay truck." That's a very nice first line for a story (it's from an old James M. Cain novel), although I've heard my neighbors do as well.

"The day before old Gilead died, he went seining in the creek," one of my neighbors told me. "And afterward, he drove the spring wagon along and gave the fish away to the neighborhood, except a sizable bass which he kept. Well, someone who must have been missed called out the warden—seining was illegal, you see—and he came out with a search warrant and found Gilead sitting under the grape arbor. The warrant wasn't any good, though, because it was his *son's* grape arbor and by the time the warden got back with a proper warrant, old Gilead had hidden the seining net in a shock of corn and slipped the bass down his overalls.

"The warden couldn't find any evidence but by this time, the barn cats had smelled fish and were standing all around Gilead. There were about fourteen of them, as I recall. The son walked the warden to his car and he got in and looked back in the yard. Old Gilead is standing there, waving goodbye to the warden, completely surrounded by cats. He is *sinking* in cats. Cats are jumping up and down, standing on one another, rubbing against Gilead's overalls. 'Them cats surely do like old Gilead,' said the warden, driving off. . . ."

It isn't the kind of story which will make the television talk shows but I like it. It has dramatic unity and certainly humor of a kind, although the listener isn't required to laugh. A story, unlike a joke, doesn't always *require* humor, rather it asks for a certain kind of *sympathy* and because it doesn't ask for laughter, it tends more often than not to get it.

The story of old Gilead also leaves the listener with a fine picture and it is from this our sympathies are naturally aroused. It is a likely and universal picture: a man, nearly caught in the act, standing on the lawn with a fish in his pants.

A Cornerstone of Pie

Family reunions, like medical checkups, usually occur annually and it is likely they do so for some of the same reasons. An examination of the lineage, spread out among the picnic tables, reveals the general health of the family.

Those of my mother's family regularly met at a country church which turned away its own members each year for the one Sunday when we came in such numbers the locals marked their doorposts the night before and kept their children inside until we had passed.

The church made this arrangement, as I recall, because my great-grandfather had built it. I never knew my great-grandfather but one of the stories whispered down was that he built a special place in the church for spitoons. He changed the architecture at the last moment, persuaded by his wife, Jessie. Great-grandfather Hart was, family legend had it, a religious man who nonetheless found sermons painful and believed a person needed all the help he could find to get through one. A chew of Mail Pouch had, after all, brought him safely to the far shores of more grievous afflictions than Methodist grandiloquence. He gave up the idea of built-in spitoons in his church but from then on he tended to refer to grandmother Jessie as "the voice from the burning spitoon."

My great-grandparents passed along their respective talents to my grandparents, he willing a mild lean toward sacrilege, and she, the ability to make a lemon meringue pie said to have weakened neighborhood marriages.

The lemon pie was for years the cornerstone of the family reunion. My grandfather always gave an artful blessing over the yawing picnic tables but he stood near the dessert table and I noticed that he took a tombstone-sized slab of the pie first, then headed off in search of fried chicken and the lesser pleasures.

Once, my cousin Pressley socked another cousin in the eye when they arrived simultaneously at the last piece, and my Uncle W. W., missing out completely, stood up on the table, a Colossus over the poor harbor of Aunt Gladys's chocolate cake, and tearfully offered two acres of prime bottomland for the last piece, no matter who had it. He said he was serious but one never knew about Uncle W. W.

Uncle W. W. had been in the Pacific during World War II, and thought all the waters of the earth belonged to it. On his first trip to Florida, he walked out and stared defiantly at the Atlantic for awhile, then returned to the house where he said, "Now you take your Pacific. *There's* an ocean. . . ." He spent most of the reunion day on the church lawn, playing croquet with the youngsters. I remember that he cheated.

I see these reunions even now as a plague of Pontiacs and Chevrolets flowing over the country to the clapboard church graced with the swelter of late summer, a sideyard of poison ivy, and enough food to have filled the larders of any undeveloped country.

They have diminished in scope and magnitude, and no longer do parchment-faced, lavender-scented ladies stand over me to say how I've grown. (My cousin Pressley, when faced once with such a scene, as we invariably were, replied, "Yes'm. When I was borned I only weighed six pounds." Whereupon his father rapped him on the head with his Masonic ring.)

The patriarch of the family is now my father, himself a grandfather, and the children are ours. We no longer meet at the country church, finding a backyard sufficient to contain us and the relatively benign spill of memories.

Being the prodigal son, I do not get back to the annual reunion often but I always think of the picture, and wish I had made the effort. I like the picture of relatives pouring over the countryside, headed for this annual conjunction, the various family constellations perfectly aligned for this one brief moment, bound by the mutual considerations of blood, proximity, and lemon meringue pie.

An Old Picture, in the Mind

When I think of the perfect community, there's an old picture in my mind. I think it is in sepia. It is the community as organism, the health of part being health of the whole. Propagation by social intercourse in the warm bed of possibility.

But the bed is colder now, and I've misplaced my soapstone. Bones choose sides in familiar churchyards, and the chicken self-destructs at family reunions. I don't harbor the illusions I once did.

The idea of the perfect community comes to me genetically, from a time when such a thing—in terms of organization and a good many personal matters, at least—did exist. But no one seemed to want to stay on.

I was talking once to a man about rural community life, and he mentioned his family's awful drive to get on to the larger places.

"What happened to those who left?" I asked.

"Some of them married women who smoked," he replied.

This is us, always an eye out for greener postures. There is an ambivalent force that drives the eye of man. It accounts for a near-universal myopia. What you see is seldom what you get. We see "community" and we get "suburb" which is another word for "enclave."

At one time in this country, communities were built by a large personal investment, one brick at a time. Care, compounded semiannually, and houses for the generations. The eye beheld then a kind of permanency.

It is therefore interesting that so few beautiful buildings have been made in the last fifty years. Or perhaps it is not interesting. The country is looking more like a movie set, fronts propped up with two-by-fours. Or: home is where the hat is. I say this because I think community is closely related to form. My prejudice is that one knows something about life in a high-rise apartment by merely regarding it.

Our architecture these days is largely immoral. Its purpose is mobility, which allows us to flee repeatedly. The modern home is Nomadic in concept, and we are the modern Bedouins, reading the want ads for split-level oases, proper benefits, and a built-in dog.

Mr. Hackney, my neighbor, says, "If you must stay and face the music, you may be inclined to write a bit of it." This is a moral but impractical viewpoint.

It is truly a practical world we inhabit. The hand is no longer related to the mind's creative moments, and men no longer live in communities. The old man down the road, nearly ninety, sits in his front yard wrapped in a blanket during the final warm days of autumn and watches the cars pass.

"People honk their horns as they pass," he says, "but no one stops. No time. No time. And I have too much."

I myself have read all the brochures about grand places. I've sniffed through the thickets of history. And I don't believe in "the perfect community." Not that something approaching it never existed, but that we chose, finally, against it. And what we have masquerading in its place is probably less satisfying, certainly more costly—yet it may be freer. We can now have almost anything. If only we could decide.

Backfield Dreams

At an early age I was determined to build a bridge for my
mother. I whirled in the backfield of early ambition, tucked
the problem securely under one arm, and turned upfield.
Where I fumbled. I was, even then, a lineman, graceful only
in small unnoticed moments but dreaming those elegant backfield dreams.

I had been instructed in grace by the poor dirt in which I
had grown up, thin and full of gristle. Farmers are not ath-
letes, you see. They grow stocky and slow from struggling to
stand upright under the burden of gravity. Or bony from running down misfortune. A man standing in such pockets of
fitful soil was never tall enough to spot his receivers. No ath-
lete stood in those mean fields (although George Thackston
used to sneak out of the back fields on his daddy's old Farmall
and drive it to high school football practice).

I'll tell anyone: I *did* mind the meanness of those fields. I
wanted riches, fame, money, the love of dark long-legged
ladies, and money. I wanted to work for myself. I was glad to
leave. I sprinted out of the barnyard, down the lane, out on
the highway, up the mountain and to school, eighty miles and
one state line away. Although the new air seemed rarified, I
was not even breathing hard.

The first thing I noticed was the rye and blue grass so thick
on the football field you couldn't feel the earth underneath.
Why, linemen would fatten here like Angus, and running backs
would founder. My daddy's whole herd of Holstein could have
wintered on that hundred yards.

The grass on high school fields I'd known had turned pale

from the heat and worn away under clumsy schoolboy cleats until there was a huge blimp-shaped patch of red clay down the center of the field. "That's so's you boys can draw plays in the dirt with a stick," our coach told visiting teams.

Over at Pickins, the whole team went in at halftime with every elbow scraped raw. Visiting teams used the girls' locker room, and the trainer lined us up by the Kotex machine and slapped one on every wound. In the third quarter a big Pickins tackle bear-hugged Cooley Long, the fullback, to the ground for a six-yard loss, then growled, "Looks like it's the wrong time of the month for you'ens."

Why, in this new-found luxuriance, grass drills and the Hamburger Drill would be a pleasure. And I would have four fine years of it. I did not know then, of course, that I would never be this innocent again.

We were merely off looking to spend the coin of our bodies' realm: energy. For what other currency do the very young have? We were not old enough to be tradesman, merchant, broker, or bailiff, nor wise enough for artist. We could not very well buy, sell, or speculate in the world's traffic. We *might* be soldiers, or ball players, both whimsical pursuits, and while a good play may have gotten you some less fortunate's ears and tail, it was, as Big Son says in James Whitehead's book, *Joiner*, short of killing and "A man's a damn fool to deny the pleasure of such moments." Football is, you see, the transformation of energy into purposeless grace. And may we not admire the manifestations of energy in nearly any form?

I don't follow the game much anymore. I never did, really. I understand, you see, that it wasn't *really* a team game. Oh, it was a *disciplined* game, alright, but the discipline merely set you free. I've lost some speed now, too, although I can still break 5.0 in the forty and my weight's still playing weight, near

198. I just tell 'em like Cooley Long used to tell 'em: "Speed ain't ever' thing, 'cause if speed *was* ever'thing, gazelles'd rule the world."

And even now, when October breaks over me like a premonition, I dream of villages meeting on green fields, wives and children eating and drinking nearby while we, the villagers, kick the skull between our boundaries. And, afterward, happy and bruised, we drink together from it.

Conversation in a Graveyard

In the graveyard, stones tilting like a drunken regiment, the old man listens through the hanging quiet as though he can hear the cedar pushing its way through Edward Gaunt's one-hundred-year-old bones. All sounds are suspect here: One might heed whispers, draw a sober breath. The old man does not say. He talks of geography but its climate is only hinted at.

In the old section of the cemetery, the rain has scoured the river rock markers, removing the resisting dates. Moss creeps into the stone-cut crevices. In the spring, the grass steals up from the bottomlands and sweeps the old, discreet markers under the green of it, diminishing tombstones to merely stone, reducing history to rumor.

The old man's cane rattles on the stones. To him, the words on the tombstones are like chapter headings in a known book. He stands slightly stooped on this island of strange land, the ponderous blooms of marble and granite twisting up around him as though he were a gardener bent by the perversity of his plants.

"Lemar," he says. "John Lemar. It could not have been so

long ago, it seems, but this granite does not lie. Fifty years, August, by these marks. Agnes Ruiner came running, saying, 'Have you heard the news?' She was breathless, a fright about her. I said, 'I reckon I have not for I do not know any.' She said, 'It's John Lemar. He's hung himself in his barn.' It was true. He fetched a cow rope and a box. He fixed the rope to his rafters and himself to the rope, then kicked the box away. There were those who figured it was his health. He had been downtown one day and remarked that he would never be a care to anyone. John Lemar. The only man I ever knew who could chew tobacco without spitting. . . ."

The stones themselves remain as silent as the bottomlands below. Archaeology will have its facts, man may think what he will, but the grave is neutral, giving only one gift: a final equality. The earth shifts from fire within and ice without; the rains fall on river rock and cornfield and then return to Anderson Fork; the grass comes in the spring.

To infrequent visitors the graveyard is always quiet. The late autumn sunlight slants obliquely through the heavy maples, and cobwebs cross the stones with their elaborate netting. The old man pushes them away with his cane. The deepening stiffness of age holds him bowed and so he moves slowly, with a slightly rocking gesture which both starts him and brings him to a stop. "I have only two speeds," he says. "They are slow and stop."

At times he pauses as one of his shoes begins to sink into the soft earth cut through by moles but he is not concerned with his own flawed gravity. He smiles at this, no conceit of his. He may be, even, *pleased.* "I have known a good woman, grew two sons to manhood. I have seen some of the country and read about much of the rest. I have lived largely a farming life and when I saw that my time was passing I did not want to

move anymore so I built a house and paid for it and it is my
home. I come to the graveyard often."

The old man turns slowly, like a cumbersome display ani-
mated by secret gears. Field mice scurry in the graveyard's
oldest corner. Leaves pile against granite. A mockingbird chat-
ters uneasily from Edward Gaunt's cedar.

"There is noise everywhere," says the old man. "Everywhere but under these stones, although I've thought under
them may be the noisiest place of all. There are such *stories*
gone with them, things no one knew or shared. My sister was
put to rest here and for a good year and a half before she went,
she heard choirs singing, in perfect harmony, the old church
songs. She was in a rest home in Xenia and she thought it was
the kind of music that is put into the walls. She worried about
it being too loud for the others, playing so late at night,
although she had no complaints. There was nothing for no
one else to hear, although I never told her. I knew what it
meant. It is a thing that has happened before to the very old
and it seems a comfort and a sign, but what is a scientific man
to make of it? I think sometimes *I* hear things others do
not. . . ."

The metaphysics of dust is not known but accepted, for
who can name what is lost in such final sleep? *Something*, says
the old man. *Something*. Memory-maddened, he is a barker
of monuments. The names spin under his cane like side-
shows. Bachelors. Old maids. Children, Suicides. And here
and there an entire family stretched from north to south, a
legion of stilled bloodlines running from stone to stone.

"There is a grave here that belongs to a Civil War soldier
named John Wesley Smith," says the old man. "In the last
year of the war, a cable came to the Smiths saying John Wesley Smith had been killed in the fighting and the body would

be shipped home. Old Mr. Smith got a team to fetch his son home but when they got it, it was no one they recognized. It was not his son, who came home on his own two feet—not laid out in a pine box. Mr. Smith brought the body home anyway and buried it decent with a rock to mark it. It was said he was a Smith from Indiana but no one came to claim him and so he stayed here.

"Over here is William Wood, barn carpenter, and on either side of him, as if related, are brothers, Edward Placethes here and Joseph Placethes there, both from Italy. They lived near the covered bridge and painted and worked about on farms and stayed to themselves. One day, Edward was found out back by an old ash hopper used for the making of soap. He had suffered a spell of some sort and it must have been severe because he had pawed the ground with his feet and dug quite a hole. It had rained on him, also. A few years later, Joseph shot himself with a pistol. No one seemed to know because it was at the time of the Xenia fair, on a Thursday, and everybody gone. Later, relatives from New York City came to claim the thirteen acres but not the body. . . ."

A man, says the old man, clings to whatever mark he can make. "When I was a boy," he says, "my father would leave his mark in the fields by his team and other farmers did the same. It was like their handwriting. They all knew, and it was enough for them. I, too, am vain about such things. I am vain about a stone. I know that it is late and foolish but we know nothing of where we are going. It is a tradition we cling to because it gives us comfort."

The old man is tired now. Hours have passed. Deep shadows fall behind the stones. "So many people I know are here. There was a building where a group of the older men would gather to sit around the stove when the weather got feisty. We spun long tales and rewrote history the way we saw it. A lot of the older ones came by, then there were not so many and one day, it was only me, alone in a large room. I thought perhaps I was early but when time passed and no one came, I knew. They were all up here, out in the yard. So I stopped going there and began to come up here because this is where my friends are. It begins there and ends here and after all these years it does not feel strange to be here anymore. Some find it foolish that I am at peace here but I ask them: What is the ground without the dead beneath it to give us comfort? I am not sure I would wish to live forever. I can hear my friends as clearly as though we were by the stove. There are a lot of them, many voices, and I feel close to them.

"Him there, I knew him since he cried. And there, I went to school with him, and worked with them, there. This one faces me, on my line, and this is my stone. Why he is on my line I don't understand but the trustees tell me they are moving it. I haven't talked to the family. Let the trustees handle it.

"A woman set out these evergreens and they grew out to here. Her husband said he would care for them and he came out with shears and trimmed them back a little but it's all over on me just the same. That vase is sitting on me but I would not call anyone on that. But why was it placed that way? The evergreens grow so fast and no one comes back except on Decoration Day maybe, and no one prunes and the stones are covered up.

"My talk is strange, I know, but I have been a few places and this is the last place I'll go and I want it to look nice. . . ."

The old graveyard has few visitors now. The regular ones are the gravedigger and the old man, who sometimes sit on the silent stones and talk. Neither thinks of himself as a visitor.

On History

Some people met at the historical society the other night to decide if they wanted to update the county history. There were no dissenters present and this brave muster of historians has since spread itself through the townships. Soon, I imagine, questions will be flying like flushed quail.

The moderator of this event, Mr. Bernard, suggested the researchers list the landmarks in each township. I was present, as a sort of neutral observer, sitting beside Mr. Doster, who has lived on an Adams Township farm almost forever, and we decided that in his township, *he* was probably the landmark.

Between the official goings on, Mr. Doster told me a little of the history of the village near him. A man who had been mistreating his wife was called to the door one night by the wife's brother. When he got there, the brother shot him, knocking him backward onto the dining-room table. "I don't mind you shooting the son-of-a-bitch," the wife said to her brother, "but he landed on my good china. . . ."

This seemed to me a good start toward documenting village life as it was lived (and not lived), but I doubt that any such stories will show up in the new history. County histories are much like small-town newspapers, a sort of public relations pageantry of local economics and pedigreed families. Actually, families used to buy their histories, like yard goods or fresh vegetables out of season. This was one way these histories were financed.

"Once in print," said Mr. Smith, a local historian, "renters suddenly owned several hundred acres of God's most favored topsoil, and fought valiantly in battles they never saw. . . ."

The writing of history, at best, is a curious business. I speak from some amount of experience, having attempted once to write a book that has in some quarters been called "history." I got some of the truth, missed a lot more, and found that some of it wasn't to be had. Truth's a bur under history's saddle: it doesn't sit well.

One of my favorite definitions of history is that it is "something that never happened, written by a man who wasn't there." No one knows who said that, which is just one more piece of testimony for the prosecution. Oscar Wilde called history gossip, Voltaire thought it was "a picture of crimes and misfortunes," and Emerson said it was "party pamphlets."

I'm a poor historian myself because I'm not much of a believer in the records. I've always preferred the more subtle recordings of literature and it seems to me that history is nothing more than second-rate literature.

I know this is aberrant thinking, and it could be merely the result of odd teachers during my youth. One semester, I had old Dr. Harley, who stood on his desk and reenacted the Battle of Verdun. The class got excited, joined in the fray, and Dr. Harley caught an eraser in his left eye from a first lieutenant on our side. His wife came along the next semester and ordered us several dozen copies of *The Carpetbaggers* by Harold Robbins, thinking it was a Reconstruction novel.

As for the writing of history, I like the idea of the man who said a historian's first duty was sacrilege. In keeping with this, I suggested to Mr. Bernard that his people include a chapter in their book on chicken thieves. I understand it's going to be a big book, over five-hundred pages, but I don't really expect to see a chicken thief in it. I wish them well anyway and I understand. History is like sinfulness; everyone has his own idea of it.

First Frost

Monday: first killing frost. Tomatoes wilted to the ground. A tentative rime of ice on the bedpost. The maple beside the back door sheds its orange leaves from the top. Goes bald, like we do.

The labrador walks morosely off to his doghouse in the corner of the woodshed. Sticks his head out to bark meagerly at whatever drives in. Watchdog services decline with the temperature.

The kitchen is too crowded. Plants, books, wood, manuscripts, vegetables. Untenable winter thoughts.

The neighborhood settles. The syrup-maker's son considers putting storm windows on the cat, takes the buck saw out of mothballs, drains the radio. The parsley wishes to sleep behind the stove.

Cat balloons with hair. Sits on the sunwarmed steps, sphinxlike. Until struck a glancing blow by a maple leaf. Then races madly around the yard leaping and chasing and dancing. One more turn, I say, before the chlorophyl drains to our root systems and we turn pale for the winter.

My neighbor fills her manure spreader with leaves. Her son comes, hitches it to the tractor, and drives across the fields, discharging the season. This month as mulch.

Story in the paper yesterday about a small child attacked by butterflies. "It was a real surprise," said the little girl. "They came out of the trees. There were perhaps a hundred. I walked through them. I felt their lips touch me. They were orange and black. Halloween butterflies, I think."

Investigation is continuing.

Bus driver passes, wiping her nose.

I'm off to dig potatoes, an act of plunder. Follow directions on an ancient piece of parchment. Find the spot, dig deep. Surprise. Potatoes roll off the pitchfork. I put them in a chest, take them to the bank. Potatoes, compounded semiannually.

Dandelions go to seed in the yard, sail away. There's corn in the highway, falling from the heavy wagons. In the neighborhood, campers fold themselves up for the season and attach to the sides of garages.

In Sabina, Martinsville, and Port William, clerks recede from their windows, sinking deeper into their shops. Cars turn over fitfully in morning garages, warming slowly to libations of anti-freeze. Restless wagonloads of wood pile past. The calendar molts. Hogs dream of fur.

The seasonal highs and lows I find personally distracting. Is there a Bureau of Complaints? Yesterday, the air was a vintage from the southern regions of France and the light was such that lepers could be healed by walking outside hatless.

My woodpile lists to the starboard. I note that it is on full. I stand beside it and snap my galluses.

Home Is Where the Hearth Is

The woodstove, pieced together by decades of tinkering fellows, is a wonderous thing, the very essence of simplicity and
straightforwardness. It is a metal box of few parts, and we are
able to pierce right to its mysterious core merely by opening a
small door and taking a look for ourselves. The emissions of a
woodstove are relatively benign; but should they, by a peculiar
downdraft or the awkwardness of the operator, emit themselves
into the living room, everyone knows immediately and the
problem is easily set right.

I once heard a newspaper editor explain to a Boy Scout
troop how a printing press worked. Knowing nothing about a
printing press allowed him to make a splendid leap. "It goes in
there, runs through here, and comes out there," he said,
pointing first to a large roll of paper, then to the perplexing
machinery. A woodstove is immanently less perplexing than a
printing press but convert the editor's brilliant reduction from
paper to a stick of firewood and the principle holds.

Its simplicity, as a matter of fact, is what got the woodstove
in trouble in the first place. This is a country built upon certain ideas, among them the dogmas of in-door plumbing and
central heating. The woodstove, sitting in America's parlors
and constantly reminding us of our humble origins, soon
clashed with the country's up-and-coming notions of sophistication. I've heard an older neighbor of mine say he thinks
we fought the Second World War for sophistication.

The woodstove manufacturers—there were only two or
three—were obviously no-nonsense people. I'd say they were
fellows who wore sensible shoes, drove black sedans and read

the hometown paper each night in front of their own stoves. Where they went wrong was failing to build a stove with chrome strips and flared fenders. Madison Avenue, meanwhile, wooed and won us with her coquetry and we were no longer a simple people. The woodstove ended up in the garage beside the Delco system and the country was inside, warming itself in front of the evening news.

Now after thirty years or so, we find history sneaking up on us again. We find ourselves, blinking like a man suddenly set down amidst a foreign population, out in the garage taking the rust off the parlor stove. This is how cycles work, of course. We know this happens. We're just never quite prepared. Mythology tells us humankind acquired fire when Prometheus stole it, later being punished by the gods. Now fire is delivered to us by the fuel oil man and we are punished at the end of the month by the bill. This is another example of how cycles work.

The stove itself, while the whimsical tunes of marketing waltz her around, keeps by and large the simple working parts Benjamin Franklin used. Many people think Mr. Franklin invented the woodstove. What he did, however, is what many inventors have done: He took a keen look at what had already been done and rearranged the various elements to his own purposes. The result was the Franklin Stove, a model of some efficiency that we still have with us.

The Franklin was my first stove and it provided ninety per cent of my heat through six Ohio winters in two drafty farmhouses. It was an adventure all the way. The Franklin is truly a parlor stove, with the capacity to heat a room or two perfectly well but it lacks the long-range military capability to take on the entire downstairs, which is what I asked it to do.

The Franklins are built much like an open fireplace, with

twin doors that open to each side. With the doors open, one may meditate delightfully upon the visible fire but the efficiency is that of a fireplace. A fireplace is generally a marvel of inefficiency. Cotton Mather tells in his diary of the ink freezing in his pen as he sat at his desk in front of the fireplace, and I've read other accounts of colonial times in which the sap forced out of burning logs turned to ice at the log ends. "You may take the poetry of an open wood fire of the present day," wrote Edward Everett Hale's mother, "but to me in those early days it was only a dismal prose."

What this means is that during the earnest part of winter, opening the doors of the Franklin is usually stark folly as the room's heat is sucked up the chimney, along with writing paper and the cat. With the doors closed, the efficiency improves to about thirty per cent (the efficiency of an open fireplace is about ten per cent) but because the doors are not airtight, the operator struggles with the factor of whimsy.

I have often thought of writing to various stove manufacturers and suggesting their stoves carry a whimsy rating, a notion to endear a manufacturer to his customers by the gift of forthrightness. Woodstove whimsy means that the stove, loaded for the night with a couple of big sticks of wood then closed down, will choose among three options: 1) if the wood is softer wood or split too thinly the fire may burn very hot all at once, then cool off as quickly; 2) if the wood is seasoned hardwood, split exactly right for the firebox, and stacked precisely in it, the fire may burn for five hours or so; 3) the fire may go out. These options qualify as adventure.

Mr. Franklin knew that whimsy was part of the process. He said that tending the stove was perhaps best suited for a man of letters, someone who would be hanging around his study at all hours of the day and able to leisurely attend the stove while reflecting upon the larger issues. Mr. Franklin built his stove in the mid-1700s, after the threat of a firewood shortage around Philadelphia. "By the aid of this saving invention," he wrote, "our posterity may warm themselves at a moderate rate, without being obliged to fetch their fuel over the Atlantic." There is, of course, a great deal of irony in Mr. Franklin's sentence, as we ponder the inscrutable nature of fuel oil delivery.

It is interesting to note that as soon as Mr. Franklin perfected his stove, he wrote an advertisement for it, warning everybody that if they continued to use the old-fashioned stoves, their teeth would go bad, their skin shrivel, and their eyes fade. Thus, he anticipated both OPEC and Madison Avenue. Mr. Franklin also organized the first volunteer fire department and was one of the country's pioneering ecologists, being the first man to turn his fireplace down.

I grew up with the woodstove occupying a prominent place in the household. I thought central heating meant exactly that: Everyone gathered around, it in the center of us. One of my favorite recollections is the night the woodstove in my grandparents' room backfired. I raced across the hall to find my grandmother crouched fearfully behind my grandfather as he, in his nightshirt, gave the stove a smart lash or two with his cane for its midnight contrariness.

After various forays into the world, where I saw some big cities and heard some big talk, I, like history, have come round again. I am back with the woodstove, by choice this time, and find it less painful than I remembered. I admit that I have succumbed to the siren call of the Scandinavian airtights and last winter was even less painful than the whimsical Franklin winter before.

The Scandinavian airtight is by and large a fine stove. The

fire does not go out when I turn my back, nor does a stick of firewood roll out and set my bathrobe on fire when I get up at half-past January for the 4 A.M. feeding. The airtight is handsomely crafted and quite a bit more efficient than the Franklin. It has taken over the living room as though the hearth were a podium and it a peroration on the energy crisis.

I understand why the ancients worshiped fire. It was provident, and it was mysterious. They didn't understand it, anymore than we understand the politics of oil, so they deified it. Our tendency, in like manner, has been to worship business, government, or science. I see using a woodstove, which has little debt to science and business and none to government, as striking a blow for a pertinent anarchy. And as for politics, I intend to vote next time for the candidate who doesn't have central heating, trusting his sermons to give off a proper heat.

The woodstove seems to be gaining converts, with no larger incentive than its implied promise of self-determination. A man I know bought a new house recently and it came with a cord of seasoned oak on the porch. He being a recent convert, the oak closed the deal. I also heard that a man in Vermont built a shopping center which he plans to heat with wood. The woodstove has actually become fashionable and any day I expect to see one on the cover of one of the newsweeklies.

I have been confirmed by the airtight stove but I keep the Franklin, an old mule in retirement pasture. It is in the kitchen, fired up on special occasions to remind me of the sense of adventure. I plan on keeping it, too. I look forward to the privilege of sounding like an Old Poot to my grandchildren, and I want to be able to point to the evidence.

A Chance of Widely Scattered Contradictions

Not having a television set, I watch the seasons for news.
Weather, outer:
Sunrise, 7:14. Character of the day, cloudy. There are
alternate swings from the ambivalence of fall to the threat of
winter. In the country, we are under the shadow of things the
pale promise of science cannot affect.
Weather, inner:
A chance of widely scattered contradictions. Thursday, I
sat under a sycamore in my shirtsleeves. Friday, I fell through
the ice. If constancy is nowhere, how do we know how to
behave? What for a model? I choose not to think of the future.
I have nothing to wear.
Amid rumors of a hard winter, I think of spring. A tractor
groaned upon the bottomlands. Inside the sealed cab rode a
transistorized young man. Loud music played, souring the
land. It could be heard for miles. Is it possible the iron plow
does poison the soil, that modern architecture *does* cause cancer? A writer I know complains that steeples and spires are no
longer built, and we do not engage the heavens. Subdivisions,
for instance, are subject to a sterner gravity. It is suggested they
do not stand well in high winds and other suburban disorders.
In spring, I begin immediately to long for the more somber
notes of fall, the stark symmetry and good taste of winter where
much if not most is left to the imagination. In summer, the
landscape becomes heavy. August swings open like an oven
door. September browns amid a dirge of cicadas. I feel conversation shrink from me. I want to immerse myself in the
language of weather and seasons. I grow curious about the soft

yellow lights in the windows of houses set back in autumn fields, and other natural phenomena.

The whole life is complicated, like the conversion of fractions. Southerners convert to Baptists by route of fear. The conversion of fractions to wholes is more difficult because fear must be expurgated. I keep potatoes in the cellar, and tomatoes late from the vine. Plants twist in the window. If I move many more times I may break in transit, like some porcelain heirloom.

Picture me so: In late fall mornings, racing over the untenable wood floors to revolve like a small planet in my irregular dancing orbit around the woodsfire, one side fiery and gaseous, the other frigid and arid, me turning upon myself spitted to this lonely, barely tillable earth.

With the advent of frost, the chickens relocate their nests, moving from the raspberry bushes to the barn. I say the egg is a gift, but that is my viewpoint. A man I know, given to obscure observation, remarks that if men could do such a thing, then guilt could be transformed into calcium and so dispelled into the straw of our circumstances and the earth would become a paradise. There is much mystery surrounding the egg. Such a mystery makes me contemplate the existence of superior beings, such as angels, prime ministers, and editors.

In October, Mr. McKay and I load our equipment in the back of his old Buick and drive rattling across the stripped cornfield to the woods. Many hours later we are one ax-handle poorer, two cords of oak richer.

The chain saw sits in the kitchen, humble now in its inti-

mations of hacking and hewing. My house is a harvest field
beckoning:

> vegetables
> porch woodpile
> unruly manuscripts needing a little off the top
> unbridled emotions to be cut down to size

The blacksmith says you do not appreciate a chain saw
until you have used an ax. The chain saw, he suggests, was
invented by a Jesuit as an instrument to test human perspective. But what fine and proper work is delivered by the ax. It
stretches the body's hidden fiber and provokes sobriety before
the fact of a two-hundred-year-old oak. With these rings (por-
traying health and disease and nature's bottomless design) I
thee wed, honest awe to as much religion as I've ever had, an
oak stump more pulpit than all the wood in a Baptist minister's
tongue. I worship here, presented with a proper mystery.

Time in its measured slide down the far side of all our
days. So much distance between points these days. The shortest line between two distances is a straight point. In this slip-
ping-down season, I am the curator of village bones. I stalk
the countryside with formaldehyde in my eyes, entombing
whatever is left. I am the bearer of tales, the keeper of the
house, the shaman shaking imagery at disease.

Preserver of tradition, I tan the hides of beasts with chicken
manure and human urine. It works wonders in my occupation
but has an adverse effect upon my social life. I have walked
from Virginia carrying a silver candelabrum and a pear tree. I
have cleared fifteen acres of hardwood with a single-bladed ax.
In February, I will spike fifteen more of sweet maple with
wooden spiles.

I will judge my existence by how many farms I have worn out. I will lose an arm in the first threshing machine in my neighborhood, will give my wife the first linoleum carpet in the village, and die of a heat stroke during grain season. There will be no mention of any of this upon my marker in the cemetery, which will read that I rested only to work.

God's Winter, Man's Lot

We've been granted a last-minute reprieve by the weather and so the countryside is filled with the sound of hatch-battening. People are seen on rooftops, peering down flue-pipes, and in backyards replacing stove bolts. Great loads of wood on wagons and trucks shuffle from woodlot to woodshed.

A solitary chimney sweep was spotted four days ago over on Williams Road, a sure sign of serious business since most people around here clean out their chimneys in the time-honored method of lowering a small, reluctant child at the end of a rope, then pulling him scuffling upward. Childless couples, or those with teenage children, usually resort to using a large chicken. Many claim this is a superior way, as the flapping of wings is said to dislodge more creosote than a wingless urchin.

A man I know swears by going to the barn and snatching several antagonistic tomcats with a fish net, throwing them in a sack with a long rope attached, then running that up and down the chimney several times.

Woodcutters are earnest men these days and I think everyone around me has a woodburning stove of some kind, or contemplates one. A fellow down the road bought one of the

newfangled makes, shaped something like a pyramid. He said
it was fine. He said he even burned a dead possum in it and
that did fine, too. "Surprising, the heat you get off a possum,"
he said.

I'm inclined to take his word for it and stick conservatively
to oak and hickory, although he suggests an unused source of
fuel, and given the way things are going I would not be surprised to one day see a classified ad for a cord of possum which
some entrepreneur gathered up from the side of the highway.

People began to get serious about woodcutting when the
energy crisis came about a few years back. I remember very
clearly when the crisis first manifested itself to Mr. McIntire.
The fuel oil man had made his customary delivery and presented the bill, which featured one of those leaps known in
track-and-field circles as "world class."

Mr. McIntire responded in the manner any normal man
would have responded. He began to jump up and down and
swear. He swore at the oil company, the Arabs and got a few
licks in at the weather, some old neighborhood animosities,
and the general maladroit nature of the cosmos itself.

"I don't know nothing about no A-rabs," said the fuel oil
man, backing off toward his truck. He failed to perceive that
Mr. McIntire's fit was of a catholic nature, aimed at restoring
symmetry to the earth, and not aimed personally at him. Mr.
McIntire's wife, feeling sorry for the fuel oil man, asked him
if he would like to have a piece of pie.

The fuel oil man, for a moment, looked as though he
would have liked a piece of pie, but pie meant walking past
Mr. McIntire, who was still jumping up and down in the yard.
The fuel oil man, by this time, was regarding Mr. McIntire as
a Bible salesman on strange turf might regard a large German

shepherd displaying his adenoids. He forgot about the pie and continued to back toward his truck. "I don't know *nothing* about no A-rabs," he said again, pulling out of the driveway.

The next day, Mr. McIntire began building a chimney. He got himself a chain saw, cut seven cords of wood, and rebuilt an old Franklin stove. He has been a woodburner ever since. I was past there a few days ago and the backyard was completely full of wood. "God can't make no winter bad enough for me," he said.

Despite the popularity of the five-day outlook, we cannot seem to unravel a pattern to the errant spin of the universe so Mr. McIntire has emerged as our preeminent philosopher. We go out to the woods these days as though it were one more crop to tend, which it is. We go out singly, in pairs and groups, and after eight winters of cutting firewood it still yields an old pleasure unconnected to economic imperative.

The main reason is that woodcutting never seems like work. In fact, I've always thought of it as a little-known sport, on the order of squash or curling. It requires something of a game plan, I suppose, but any number can play and referees and timekeepers aren't usually necessary. Its major drawback is that it isn't much of a spectator game, as the woodlot pleasures are mostly individual.

It's becoming more popular with participants, however, and suddenly big money has come to the woodlot. I hear that a cord of hardwood fetches a hundred and fifty dollars along the eastern seaboard. A man I know up in Chagrin Falls says he's ashamed of himself but he's been getting a hundred and five dollars. If the trend continues, folks may begin keeping firewood in a safety deposit box. Around here, a cord is still a reasonable purchase and most of the serious converts cut their own. One fellow I know declared his household was going to be heated with wood and to make sure no one sneaked out and turned up the thermostat, he pulled the breakers to the furnace and hid them in the barn.

Wood was once a primary heat source and about the time of the Civil War the average American family burned 17.5 cords of it each year. For quite some time after that, it was a common practice out in the provinces to pay the minister a little cash, and his wood supply. I don't know how the minister feels about such an arrangement these days but firewood seems to be back as a desirable exchange and perhaps soon we'll be seeing firewood quotations next to money funds in the newspapers.

When I go after firewood, I prefer to go with someone such as Mr. McIntire because he is both good company and a craftsman. Sometimes he is both at the same time. Recently, we spent half an hour maneuvering his truck delicately around a stand of young maple saplings to get at a downed hickory. "I expect," said Mr. McIntire, "that Confuscius has a saying about this. . . ."

More times than not, I go out alone, arising just before dawn and making tea while fooling with the stove, which is cast iron and forbearant, yielding slowly to my impatiences. I regard the out-of-doors through the kitchen windows, looking for stars above the backyard maple. The old labrador stretches on the back porch and puts his head in the food dish, banging it around although he knows he has not missed anything. He never misses the sound of food, and can hear a biscuit drop on the grass a hundred yards off.

When I reach for my woods' boots, he gets excited, torn between the pleasures of breakfast and getting to the woods, a dilemma that divides him equitably between dish and door. He bolts a cup of dry food, then pushes the porch door open with his head and runs to the old Volkswagen, which has the back seat out to make room for him, the chain saw, an ax, and other woodlot paraphernalia.

By first light we are driving into the middle of the woods on the old railroad bed. Because it is graded, it stays dryer longer than the woods' floor and I use it to bring out firewood. It is a mild irony that the old line still hauls a load of freight from time to time. People still refer to it as the Grasshopper because, they say, it was so slow grasshoppers never bothered jumping off the roadbed. When it ran through these woods, passengers shot pheasant out the windows and, without the train slowing, jumped off and retrieved them.

The line ran between the communities on either side of me, and a few miles to the northeast made a connection to Cincinnati and thus the world. Elmer Lemar, who lives down the road from me, told me his father, Black John Lemar, got on it one day to see where he might go, found himself in a Cincinnati saloon, and under the influence of barley-provoked bravado, joined the army. The army sent him out west where he was in the unit supposed to relieve Custer at Little Big Horn. Black John, finally arriving back home after an absence of several years, told the neighbors that if the Grasshopper had been any faster he'd have returned wearing a toupee woven out of broom sedge.

The only other historical note to the Grasshopper is that it was once owned by Henry Ford. A local historian reports Mr. Ford ran it with his customary attention to detail; while the crew waited for loading and unloading at the Kingman switch, they were kept busy by polishing the engine. Mr. Ford's future, unfortunately, did not lie with the locomotive and so we gained the automobile but lost the Grasshopper.

The historical persona I consider in the woods is not Henry Ford, however, but old Sam Ellis, who cut down a tree on himself near here. They amputated Sam's leg in the woods and later buried it over at the Methodist church, in a baby's casket.

Old Sam recovered, and lived on for some time. He had a wooden leg and used thumbtacks to hold up his socks. His grandchildren sometimes tried using tacks on their socks, too, but found it painfully unworkable. When old Sam died, they buried the rest of him beside his leg.

Thinking of Sam keeps me honest in the woods, reminding me of the caprices in going after firewood. Of course, there are caprices either way. Caprice of fuel oil bill, caprice of chain saw and falling timber. I take my chances, a mental historical marker to old Sam in my mind. The sources of all energy are mysterious but the power of the sun, locked simply into a stick of firewood, seems a benign enough technology. I can understand most of it myself.

Inching Up on Winter

Here in the country we inch up on winter, as though it were our idea. This illusion is one of Ohio's more fetching traits, this willingness to transport us reasonably from one season to another without those sudden ascents and descents that produce bends of the spirit.

Winter around my place scorns December 22 as its official

port-of-entry and, like unwanted relatives, may drop in any-
time between October and the end of the year.

It is never much of a surprise, though, because there's been
plenty of warning. The backyard maple sheds its leaves from
the top, causing an eight-year-old down the road to say, "That
tree looks like it's standing in its underpants." The cat's coat
thickens and she takes to sneaking into the doghouse at night,
turning the old labrador up to five.

The light seems to thin through October and into November, losing that mid-autumn richness. It is the look of light
only certain Flemish painters managed to put down. Sounds
change, too. Doors shut. Storm windows go up. There is a
creaking in the fields, the sound of unoiled, high-pitched
machinery: a flock of starlings, barely of God's own, gleaning
the cornfields.

The rainy days come. The energetic music of early autumn
shifts to something slow and impeding. We look for blue skies
in the encyclopedia.

The first day of winter is the first day the mercury straggles
up, just missing forty-five degrees under a sky like old laundry.
It is not that we haven't been warned, merely that we are unaccustomed. To be unaccustomed is probably a mark of the species, since we lack the ears of the coyote and the nose of the
possum.

There is a difference between being unaccustomed, however, and being surprised. Surprise is not a modern characteristic, I think, because it is a bit naïve and likely seen as
something to use to one's advantage. Surprise is an old trait,
like manners or good humor, mostly found these days in attic
kegs and the bottom drawers of bureaus in spare rooms.

When the clouds break, the pale sunlight spreads subtly

over patches of cornfield here and there in the distance, making the heart lift. And here's the real crisis: will there be enough sunlit moments, no matter how pale, between now and spring?

First Snow, Baleful Mercury

The first big snow of the season is upon us and there seems to be a shoulder-shrugging sang-froid loose in the neighborhood. Down the road I saw a man standing outside, up to his premises in snow and baleful mercury, smoking his pipe. As I recall, this is the same fellow who last year at this time threatened to hijack a snowplow to take him and his wife to St. Petersburg.

There was, actually, a snowplow hijacking in the county last year. At the height of the blizzard, the police chief of a nearby village took over the village snowplow, ensconced a badly pregnant lady in the cab, and plowed her to a Cincinnati hospital. The chief, in the better traditions of old western Golly-m'am modesty, later said, "All in a day's work," as though modern police work included pregnancy as just one more kind of misdemeanor to be reckoned with.

The attitude up and down the road this year does seem to be an old-west one, I think. This year, nobody's stepping off the sidewalk to let winter pass. We're winter's sheepherders, finally on our feet and demanding grazing rights.

As with most things, there's two sides to such thinking. We've gotten feisty and that's all right. There's character in it. But I expect our attitude toward snow as something malevolent, to suffer under, is an impatient and contemporary one. It likely comes in part from the way we have increasingly come to view the outside world as a sort of twenty-four-hour super-market, and access as part of our inheritance.

It's true that today all roads lead not to Rome but to the supermarket and I don't doubt that future archaeologists, sifting the ruins of shopping centers, will put together Twentieth Century Man as an ambivalent creature caught between the contradictions of canned goods and fresh produce.

Well, we're a strong-arm, impatient people, and it's our trait not only to overcome but to overwhelm. A man with stock in snowmobiles, snowblowers, or a four-wheel drive anything is surely a man whose fortunes are on the rise. And I'll admit here to my own sporadic craving to bare-knuckle this damned, pointed month.

The character I admire, however, is a Vermont farmer Scott Nearing talked about once. He had a place near the New York State line and after a survey, the engineers found he wasn't in Vermont after all; he was in New York. So they went out and informed the old fellow.

"Thank God," he said, upon hearing the news. "I don't think I coulda stood another one of them Vermont winters. . . ."

A Christmas Memory

My grammar school was condemned. Does that explain why I was a backward child, unable to spell out my desires at Christmastime? I hated the linoleum on the floor of my unheated upstairs room because on winter mornings when my feet touched it I thought my heart would stop and so I wished for carpeting. "What's carpeting for a child to ask at Christmas?" said my father. I learned carpeting was adult, and so desire persisted, having nothing to do with carpeting anyway.

I thought I wanted Mary Frances Verdin, too, but that could have been because Toby Hunter had her. Toby Hunter's house had central heating and Toby Hunter had Mary Frances Verdin. That was the way I saw it.

I wanted, of course, what I could get, but that was secondary. Anything I could name I supposed I might be able to get. Even though I wanted Mary Frances Verdin, I was careful not to *pray* for her.

If anyone in the family had read, they would have given me books. Only my grandmother read, however, and very carefully, as though reading were a barefoot act in a rocky field. For Christmas, someone gave her a hardbound copy of *Catcher in the Rye*. She found that Holden Caulfield took liberties with the language, turned to the back to see if he talked any better and when he didn't, she burned the book.

We cut the tree on the first Saturday after school was out. I was fourteen, the oldest, and so in charge. My two brothers went with me into the cedars on the hillside pasture. I was looking for a symmetry which, of course, I could never quite find. Because it was Christmastime, my brothers deferred to my impossible standards. After an hour or more, still unsatisfied, I made my choice.

The tree went in the front room, which was never used except for special occasions, and the Christmas lights lit the gloomy front of the farmhouse. Uncle W. W. fell into it once, after he had made too many trips to the laundry hamper. He kept a bottle of whisky under the dirty clothes at Christmastime. Both the tree and Uncle W. W. recovered, although it was touch-and-go with my grandmother.

Grandfather never drank but once, a fact which spoiled my grandmother. On a trip to New York, my grandparents went to a night club because they had never been to one. My grandfather ordered a mint julip. It was served in a tall frosted glass with a large clump of mint and an exorbitant bill. "The goddamned shrubbery is sure high out here," said my grandfather.

Half the family was present, my mother's side, twenty-five in the farmhouse on Christmas Day, mamma's bed sagging under coats, family gifts piled in tiers around the tree, the laundry hamper armed. We spread out to keep the house from listing.

When we went in for dinner, a cat was sitting on the linen cloth lapping gravy from a china bowl. Father picked up a hammer on the pantry shelf, tossed it at the cat, struck him on the head and he leaped to fall dead on the floor, the hammer sailing into a chair, not a dish broken, grandfather saying, "Praise God for eyesight to protect a man's hearth and gravy!" And we ate.

There are moments when we are young, and all of time seems neatly balanced. Past, present, future, all aligned, as though the machinery of the universe had inexplicably hesitated, allowing me sight that was curved, like time itself was said to be. It was after dinner and I was lying behind my grandfather's chair, to one side of the fireplace. The voices in the room blended into one pleasant droning sound.

I imagined my grandfather's life, occurring to me then as a series of images accompanied by sensations that unreeled in my mind like a film speeded up. I saw myself in the pause, the waiting time of my life, and then I saw myself as my own grandfather, more a sensation of how I wished to be than a picture.

Desire, uncalibrated, unknown, moved me, and I knew I wanted everything, or nothing, though I could not name anything. O, I thought, almost in pain, for it was what I *could* name, for my parents, grandparents, uncountable sweetly sweating cousins, fat uncles, all the Christmas lights, warmth, and pleasures of this room forever! Amen.